SUPERCRITICAL WATER

SUPERCRITICAL WATER

A Green Solvent: Properties and Uses

Yizhak Marcus

A JOHN WILEY & SONS, INC., PUBLICATION

Published by John Wiley & Sons, Inc., Hoboken, New Jersey
Published simultaneously in Canada

For general information on our other products and services or for technical support, please contact our Customer Care Department within the United States at (800) 762-2974, outside the United States at (317) 572-3993 or fax (317) 572-4002.

Wiley also publishes its books in a variety of electronic formats. Some content that appears in print may not be available in electronic formats. For more information about Wiley products, visit our web site at www.wiley.com.

Library of Congress Cataloging-in-Publication Data:

Marcus, Y.
  Supercritical water : a green solvent, properties and uses / Yizhak Marcus.
     p. cm.
  Includes indexes.
  ISBN 978-0-470-88947-3 (hardback)
  1. Solvents. 2. Green technology. I. Title.
  TP247.5.M294 2012
  541'.3482–dc23

                                                                2011049802

10 9 8 7 6 5 4 3 2 1

CONTENTS

Preface   ix

List of Acronyms and Symbols   xiii

1   Introduction   1

    1.1   Phase Diagrams of Single Fluids / 1
    1.2   The Critical Point / 3
    1.3   Supercritical Fluids as Solvents / 5
    1.4   Gaseous and Liquid Water / 8
    1.5   Near-Critical Water / 15
    1.6   Summary / 17

2   Bulk Properties of SCW   22

    2.1   Equations of State(EoS) / 22
        2.1.1   PVT Data for SCW / 22
        2.1.2   Classical Equations of State of SCW / 24
        2.1.3   Scaling Equations of State for SCW / 26
        2.1.4   EoS of Supercritical Heavy Water / 29
    2.2   Thermophysical Properties of SCW / 30
        2.2.1   Heat Capacity / 30
        2.2.2   Enthalpy and Entropy / 32
        2.2.3   Sound Velocity / 34
    2.3   Electrical and Optical Properties / 34
        2.3.1   Static Relative Permittivity / 34
        2.3.2   Electrical Conductivity / 37
        2.3.3   Light Refraction / 38
    2.4   Transport Properties / 39
        2.4.1   Viscosity / 39
        2.4.2   Self-Diffusion / 41

2.4.3   Thermal Conductivity   /   42
2.5   Ionic Dissociation of SCW   /   44
2.6   Properties Related to the Solvent Power of SCW   /   47
2.7   Summary   /   49

3   Molecular Properties of SCW   57

3.1   Diffraction Studies of SCW Structure   /   60
    3.1.1   X-Ray Diffraction Studies of SCW Structure   /   61
    3.1.2   Neutron Diffraction Studies of SCW Structure   /   62
3.2   Computer Simulations of SCW   /   66
    3.2.1   Monte Carlo Simulations   /   67
    3.2.2   Molecular Dynamics Simulations   /   70
3.3   Spectroscopic Studies of SCW   /   74
    3.3.1   Infrared Absorption Spectroscopy   /   74
    3.3.2   Raman Scattering Spectroscopy   /   77
    3.3.3   Nuclear Magnetic Resonance   /   79
    3.3.4   Dielectric Relaxation Spectroscopy   /   82
3.4   The Extent of Hydrogen Bonding in SCW   /   83
3.5   The Dynamics of Water Molecules in SCW   /   90
3.6   Summary   /   92

4   SCW as a "Green" Solvent   100

4.1   Solutions of Gases in SCW   /   101
    4.1.1   Phase Equilibria   /   101
    4.1.2   Interactions in the Solutions   /   104
4.2   Solutions of Organic Substances in SCW   /   106
    4.2.1   Phase Equilibria   /   106
    4.2.2   Interactions in the Solutions   /   111
4.3   Solutions of Salts and Ions in SCW   /   115
    4.3.1   Solubilities of Salts and Electrolytes   /   115
    4.3.2   Thermodynamic Properties   /   121
    4.3.3   Transport Properties   /   123
    4.3.4   Ion Association in SCW   /   129
    4.3.5   Ion Hydration in SCW   /   134
4.4   Binary Mixtures of Cosolvents with SCW   /   138
4.5   Summary   /   141

5   Applications of SCW   151

5.1   Conversion of Organic Substances to Fuel   /   152
    5.1.1   Conversion to Hydrogen and Natural Gas   /   152
    5.1.2   Conversion to Liquid Fuel   /   156
5.2   Supercritical Water Oxidation   /   157

5.2.1 General Aspects of SCWO Process / 158
5.2.2 Examples of SCWO Applications / 160
5.3 Uses of SCW in Organic Synthesis / 162
5.4 Uses in Powder Technology of Inorganic Substances / 164
5.5 Geothermal Aspects of SCW / 166
5.6 Application of SCW in Nuclear Reactors / 169
5.7 Corrosion Problems with SCW / 171
5.8 Summary / 174

Author Index   183
Subject Index   199

PREFACE

As of the summer 2011, there were more than 3000 topics dealing in detail with supercritical water (SCW) in the SciFinder literature search instrument of the American Chemical Society. However, there were more than 14,000 entries outlining this concept. In the 1980s some 100 papers and in the 1990s some 900 papers on supercritical water were published, while at present there are already more than 2000 papers. As it is impossible to compile all the published information in a book, an attempt has been made to include the maximum possible important properties and uses of supercritical water. Factual information is given in numerous tables along with suggested references for more details on the subject. Where appropriate, the reader is referred to several reviews relevant to the topics included in this book.

Prior to 1980, only a few dozen papers dealt with SCW, considering SCW mainly within the broad subject of high-pressure steam in the context of electric power generation. The papers dealt principally with the heat transfer in SCW, mineral solubilities in it, and corrosion by it. E. U. Franck, a pioneer in the study of supercritical fluids, however, published in 1968 a review (*Endeavour*, **22**, 55) that highlighted some of the properties of this fluid and its possible uses. The properties contrasted with those of water vapor and of the liquid water at ambient condition. They included the complete miscibility of SCW with nonpolar fluids, the very high mobility of ions from electrolytes dissolved in SCW, and the water itself acquiring appreciable electrical conductivity. Knowledge of the chemical behavior of high-temperature water, in the pure state and when serving as a solvent, led to the understanding of the structural features of SCW and of hydration phenomena in it. The properties of geochemically important "hydrothermal" solutions could also be explained and possible technical applications were suggested.

The properties of dense steam or compressed hot water below the critical point and solutions in such media can be of interest, as these are able to act as "green" solvents. In the present book, the so-called near-critical water is however only cursorily dealt with, as it is mainly devoted to the properties and uses of supercritical water. SCW in itself can also be deemed to be a "green" solvent, that is, environment-friendly.

Having dealt for many years with liquids and solutions, the author's interest in SCW was raised by the proposal he received in the late 1990s from the INTAS agency for his participation in an international collaboration on this subject. During the period of 3 years of the project that involved three groups from Russia, one from Greece, one from Germany, and the present author, "experimental and theoretical studies of supercritical aqueous solutions as a medium for new environmentally friendly and energy efficient technologies of pollution control" were carried out. As a further result of this collaboration, one of the participants, A. Kalinichev, was invited together with the present author by R. Ludwig, the editor of the book *Water: From Hydrogen Bonding to Dynamics and Structure*, to write a chapter on SCW for it. This provided the initial impetus to the writing of the present book, seeing that none existed so far that summarized the state of the art and in view of the increasing interest in the subject as reflected by the increasing number of published papers.

The book is divided into five chapters. Chapter 1 introduces supercritical fluids in terms of the phase diagrams of the fluids and their critical points. A brief description of a variety of supercritical fluids that have been used as solvents is given. Attention is then turned toward the water substance, in its gaseous state (water vapor) and ordinary liquid water and their properties. As water is heated toward the critical point, near-critical water is reached, and a short discussion of this state of water (that has found some applications as a "green" solvent) is presented. Chapter 2 deals with the macroscopic measurable properties. Foremost of these are the temperature–pressure–density or volume (*PVT*) relationships described by means of equations of state. Other important thermophysical properties of SCW are the heat capacity and the enthalpy and entropy. The electrical and optical properties include the static dielectric constant, the light refraction, and the electrical conductivity of neat SCW. The transport properties involve the viscosity, the self-diffusion, and the thermal conductivity. The ionic dissociation of SCW is then discussed, and finally the properties of SCW relevant to the solubility of solutes in it are briefly described. Chapter 3 deals with the structure and dynamics as inferred from experimental data and computer simulations. Diffraction of X-rays and in particular of neutrons provides information on the molecular structure of SCW. Computer simulations provide information on both the structure (by the Monte Carlo method) and the dynamics (the molecular dynamics method). Spectroscopic studies, involving infrared light absorption, Raman light scattering, nuclear magnetic resonance, and dielectric relaxation, complement the aforementioned studies. It is shown that SCW has appreciable hydrogen bonding between its molecules and the extent of this is explored. Finally, the dynamics of the water molecules in SCW and the lifetime of various configurations in it are discussed. Chapter 4 describes the solubilities of gases, organic substances, salts, and ions in SCW in terms of the relevant phase equilibria. The interactions that take place between the water molecules and the solutes of the various categories are presented. In particular, for ions and salts, the properties of such solutions are dealt with. In case of ions, their association on the one hand and their hydration on the other determine these properties. Finally, Chapter 5 includes the current practical uses, whether on a modest or on a full industrial scale. Conversion of biomass to fuel, gaseous or liquid, is one such use. SCW oxidation (SCWO) of pollutants and hazardous materials is another important use, the problems associated with which have not so far been completely resolved. Some other uses include organic synthesis, where SCW is both a reaction medium and a reactant, nanoparticle production of inorganic substances (mainly oxides), and as a neutron moderator in nuclear power reactors and at the same time as the coolant, providing the fluid for turbine operation. Geochemistry is another field where SCW plays a role, because deep strata in the earth's crust provide the high temperature and pressure to convert any water derived from hydrous minerals to SCW. This is then evolved in thermal vents, carrying along some minerals dissolved in it. Finally, some of the corrosion problems met with in applications of SCW are briefly dealt with.

A vast amount of information is available on SCW and solutions therein; this book however provides those numerical data in tables that help the reader to appreciate the quantitative aspects of SCW and its properties. Some other tables include annotated examples of the uses of SCW, but on the whole, it is possible only to point out what various authors have studied, to summarize it, and as necessary to comment on this. This book includes references to which the readers interested in having in-depth knowledge of the topics may refer. It is hoped that the book will help understand the concept of supercritical water, its properties, and uses.

YIZHAK MARCUS

Jerusalem, 2011

LIST OF ACRONYMS AND SYMBOLS

ACRONYMS

EoS	Equation of states
MC	Monte Carlo computer simulation
MD	Molecular dynamics computer simulation
PVT	Pressure–volume–temperature
SAFT	Statistical associated fluid theory
SCD	Supercritical carbon dioxide
SCF	Supercritical fluid
SCW	Supercritical water
SCWG	Supercritical water gasification
SCWO	Supercritical water oxidation
VLE	Vapor/liquid equilibrium

SYMBOLS

Symbol	Description	Units
Universal constants		
e	Unit electrical charge: 1.602177×10^{-19}	C
F	Faraday's constant: 9.64853×10^{4}	$C\,mol^{-1}$
k_B	Boltzmann's constant: 1.380658×10^{-23}	$J\,K^{-1}$
N_A	Avogadro's number: 6.022136×10^{23}	mol^{-1}
R	Gas constant: 8.31451	$J\,K^{-1}\,mol^{-1}$
ε_0	Permittivity of vacuum: 8.854188×10^{-12}	$C^2\,J^{-1}\,m^{-1}$
Physical quantities		
A	Helmholtz energy, molar	$kJ\,mol^{-1}$
a	Attractive parameter in EoS	$J^2\,Pa^{-1}\,mol^{-2}$
B	Virial coefficient	m^3

Symbol	Description	Units
b	Scattering length	nm
b	Covolume parameter in EoS	m^3
Co	Number of components	
C_P	Heat capacity at constant pressure, molar	$J\,K^{-1}\,mol^{-1}$
C_V	Heat capacity at constant volume, molar	$J\,K^{-1}\,mol^{-1}$
c	Concentration, molar scale	$mol\,dm^{-3}$
D	Diffusion coefficient	$m^2\,s^{-1}$
Df	Number of degrees of freedom	
d	Interatomic distance	nm
E	Energy, molar	$kJ\,mol^{-1}$
e_{HB}	Energy of an H-bond	$kJ\,mol^{-1}$
f	Fugacity	kPa
f_i	Fraction of water molecules with i H-bonds	
G	Gibbs energy, molar	$kJ\,mol^{-1}$
g	Kirkwood dipole orientation parameter	
$g(r)$	Pair correlation function	
H	Enthalpy, molar	$kJ\,mol^{-1}$
h_i	hydration number, ionic	
I	Ionic strength	$mol\,dm^{-3}$
I	Intensity of scattered or absorbed radiation	
K	Equilibrium constant	$dm^3\,mol^{-1}$
K_W^O	Octanol/water partition coefficient	
K_W	Ion product of water	$dm^6\,mol^{-2}$
k	Variable, function of diffraction angle	m^{-1}
k	Rate constant	s^{-1}
L_g	Ostwald coefficient of gas solubility	
M	Mass, molar	$kg\,mol^{-1}$
m	Mass, molecular	kg
m	Molality	$mol\,kg^{-1}$
N	Number of components	
N	Number of molecules or ions	
N_{co}	Coordination number	
n	Refractive index (at specified frequency)	
$\langle n_{HB}\rangle$	Average number of H-bonds per water molecule	
P	Pressure	MPa
Ph	Number of coexisting phases	
P_i	Internal pressure	MPa
p	Vapor pressure	kPa
p (event)	Probability of an event	
r	Radial distance from a particle	nm
r_i	Radius, ionic	nm
S	Entropy, molar	$J\,K^{-1}\,mol^{-1}$
$S(k)$	Structure factor	
s	Solubility	$mol\,dm^{-3}$

Symbol	Description	Units
T	Temperature (absolute)	K
T_1	Spin-lattice NMR relaxation time	s
t	Temperature	°C
t_i	Transference number, ionic	
U	Configurational energy of system, molar	$kJ\,mol^{-1}$
u	Sound velocity	$m\,s^{-1}$
u_i	Mobility, ionic	$m\,s^{-1}\,V^{-1}$
V	Volume, molar	m^3
v	Specific volume	$m^3\,kg^{-1}$
v	Velocity, molecular	$m\,s^{-1}$
x	Mole fraction of specified component or species	
Y	Generalized thermodynamic function	
y	Activity coefficient, molar scale	
Z	Compressibility factor PV/RT	
Z	Lattice parameter	
z	Charge number, algebraic	
α	Kamlet–Taft H-bond donation ability index	
α	Polarizability, molecular	m^{-3}
α_p	Expansibility, isobaric	K^{-1}
β	Kamlet–Taft H-bond acceptance ability index	
γ	Activity coefficient, molal scale	
δ	NMR chemical shift	ppm
δ_H	Hildebrand solubility parameter	$MPa^{1/2}$
ε_r	Relative permittivity	
ζ	Reciprocal of the friction coefficient	
η	Dynamic viscosity	$mPa\,s^{-1}$
θ	Angle between two dipoles or three atoms	°
κ_S	Compressibility, adiabatic	GPa^{-1}
κ_T	Compressibility, isothermal	GPa^{-1}
Λ	Conductivity, molar	$\Omega^{-1}\,m^2\,mol^{-1}$
λ_i	Conductivity, ionic equivalent	$\Omega^{-1}\,m^2\,mol^{-1}$
λ_{th}	Conductivity, thermal	$W\,K^{-1}\,m^{-1}$
μ	Dipole moment	$C\,m$
μ	Chemical potential	$kJ\,mol^{-1}$
v	Wave number	cm^{-1}
ξ	Correlation length	m
π^*	Kamlet–Taft polarity/polarizability index	
ρ	Density	$kg\,m^{-3}$
σ	Diameter, molecular	nm
τ	Relaxation or correlation time	ps
φ	Volume fraction	
χ	Number of H-bonds of any given water molecule	
ω	Frequency	s^{-1}
ω	Pitzer's acentric factor	

SUBSCRIPTS AND SUPERSCRIPTS

$*$	Pure substance or scaling parameter
o	Standard state
∞	Infinite frequency
∞	Infinite dilution
ass	Association
c	Critical
D	Dielectric or Debye, or self-diffusion
E	Excess thermodynamic quantity
f	Formation (thermodynamic quantity)
g	Gas
ig	Ideal gas
r	Reduced, divided by the critical quantity
v	Vaporization
w	Water
σ	Saturation (VLE equilibrium)

1

INTRODUCTION

1.1 PHASE DIAGRAMS OF SINGLE FLUIDS

Substances appear in Nature in several states of aggregation: crystalline solids, amorphous solids, glasses, liquids, and gases. The latter two states are collectively termed "fluids" because they are subject to flow under moderate stresses (forces). A phase is a portion of space in which all the properties are homogeneous, that is, they do not depend on the precise location in the phase (except at its boundaries). Depending on the external conditions (thermodynamic states) of temperature and pressure, a substance may exist at several states of aggregation at the same time, each of which is represented as a different phase. These phases may be at equilibrium with each other, and the phase diagram, in terms of the external conditions of the temperature and pressure, represent which phases are at equilibrium with each other. These phase equilibria have several important features, governed by the phase rule of Gibbs. This rule states that number of degrees of freedom (Df) in a system is equal to 2 plus the number of components (Co) minus the number of phases that exist at equilibrium (Ph):

$$Df = 2 + Co - Ph \tag{1.1}$$

The degrees of freedom of interest in the present context are the external conditions that can be independently chosen: the pressure P, the temperature T,

Supercritical Water A Green Solvent: Properties and Uses, First Edition. Yizhak Marcus.
© 2012 John Wiley & Sons, Inc. Published 2012 by John Wiley & Sons, Inc.

and the composition of mixtures. When the latter are expressed as the mole fractions of the components: $x_1, x_2, \ldots, x_N$ for an N-component mixture, there are $N - 1$ independent composition variables. A component is a substance that can be added independently: water is an example of a component and a salt, such as NaCl, is another example, but each of their constituent ions (H^+ or H_3O^+, OH^-, Na^+, Cl^-) is not. Only neutral combinations of the ions can be considered as components, since they can be actually handled.

For a single substance $Co = 1$ and up to three phases can exist at equilibrium, in which case $Df = 0$, there remain no freely determinable external conditions (no degrees of freedom): the temperature and pressure are fixed. This invariant point (involving a solid, a liquid, and their vapor) is called the triple point of the substance. In the case of water: ice, liquid water, and water vapor are at equilibrium at $0.01°C$ ($T = 273.16$ K) and $P = 0.61166$ kPa [1].

Two phases of a single component at equilibrium permit according to the phase rule (1.1) a single degree of freedom: either the temperature is variable and the pressure is then fixed or for a variable pressure the temperature is fixed. This simple function determines a line in the two-dimensional phase diagram. For the water substance, at temperatures below freezing several phases of ice exist, depending on the pressure, and a pair of which can exist at equilibrium along the lines of the relevant phase diagrams. Ice may sublime to form water vapor, and again the two phases may exist at equilibrium along the sublimation line of the phase diagram of water. These aspects of the phase diagram of water are outside the scope of this book. The phase diagram of water is shown in Fig. 1.1, the phase boundaries being the coexistence lines of two phases at equilibrium. The phase diagram of water at very low temperatures and very high pressures is complicated by the existence of several ice phases of different densities and structures (not shown in Fig. 1.1), but these are of no concern in the present context.

When attention is drawn toward fluid phases, vapor–liquid equilibria (VLE) constitute a very important subject. Again, in such two-phase systems the feature of the phase diagram of water is a line, called the saturation line for VLE, and is designated by the subscript $(_\sigma)$. This line for water is shown in Fig. 1.1, extending from the triple point up to the critical point (see below). The fact that the normal boiling point of water at ambient pressure (1 atm $= 1.01325$ bar $= 101.325$ kPa) happens to correspond to $100°C$ (373.15 K), is incidental in this respect. Wagner and Pruss reported the IAPWS 1995 formulation for the thermodynamic properties of ordinary water substance [1] that appears still to be the last word on the subject. The expression for the saturation vapor pressure $p_\sigma(T)$ takes the following form:

$$\ln(p_\sigma/P_c) = (1 - \tau)^{-1}[a_1\tau + a_2\tau^{1.5} + a_3\tau^3 + a_4\tau^{3.5} + a_5\tau^4 + a_6\tau^{7.5}] \quad (1.2)$$

TABLE 1.1   Parameters for the Saturation Vapor Pressure of Water, Eq. (1.2)

P_c/MPa	T_c/K	a_1	a_2	a_3	a_4	a_5	a_6
22.064	647.096	−7.8595	1.84408	−11.78665	22.68074	−15.96187	1.801225

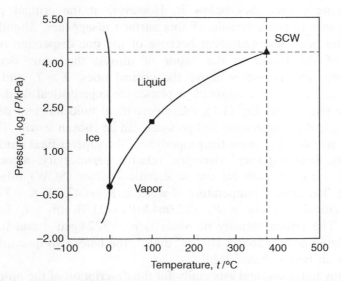

FIGURE 1.1   The phase diagram of water: point • corresponds to the triple point ($t_{tr} = 0.01°C$, $P_{tr} = 0.611\,kPa$), point ▼ to the melting point ($t_m = 0°C$, $P_m = 101.3\,kPa$), point ■ to the normal boiling point ($t_b = 100°C$, $P_b = 101.3\,kPa$), and point ▲ to the critical ($t_c = 374°C$, $P_c = 22.1\,MPa$). The line between • and ▲ is the saturation line, $P = p_\sigma$. Various compressed ices exist above point ▼ that are of no concern in the present context.

where P_c is the critical pressure and $\tau = 1 - T/T_c$, with T_c being the critical temperature. The values of P_c, T_c, and the coefficients a_i ($i = 1–6$) are shown in Table 1.1.

Numerical values of p_σ at several temperatures are shown in Table 1.5 along with data for other properties of liquid water discussed in Section 1.4.

A supercritical fluid (SCF) consists of a single phase and since $Co = Ph = 1$ it has two degrees of freedom according to Eq. (1.1). Its temperature and the pressure can be chosen at will, provided they are larger than the critical values (see below).

1.2   THE CRITICAL POINT

As the temperature and pressure of a fluid increase, a point is reached where the two phases, the liquid and the vapor, coalesce into a single phase.

The density of the liquid diminishes along the saturation line whereas that of the vapor increases as its pressure increases, until finally they become equal at the critical point. This is characterized by a critical temperature T_c, a critical pressure P_c, and a critical density ρ_c. A liquid that is confined in a vessel in a gravitational field has a free surface with respect to its vapor, hence, being denser, lies below it. However, at the critical point the meniscus that is characteristic of this surface disappears. Slightly below T_c the fluid becomes opalescent because of the fine dispersion of minute droplets of the liquid in the vapor of almost the same density. At temperatures and pressures above the critical ones, $T > T_c$ and $P > P_c$, the substance exists as a single clear phase, the supercritical fluid. According to the phase rule, Eq. (1.1), the supercritical fluid has two degrees of freedom, and the temperature and pressure can be chosen at will. These two external variables determine the properties of the supercritical fluid, such as its density, heat capacity, viscosity, relative permittivity, among many others, as are dealt with for the supercritical water (SCW) substance in Chapter 2. The critical temperature of water is $T_c = 647.096\,\text{K} = 373.946°\text{C}$ and its critical pressure is $P_c = 22.064\,\text{MPa} = 21.78\,\text{atm}$ (cf. Tables 1.1 and 1.3). The critical density of water is $\rho_c = 322\,\text{kg m}^{-3}$ and its critical molar volume is $V_c = 56.0 \times 10^{-6}\,\text{m}^3\,\text{mol}^{-1}$, the latter two quantities being known to no better than $\pm 1\%$.

For many purposes, and especially for the description of the properties of supercritical fluids, it is expedient to use the reduced variables: the reduced temperature, $T_r = T/T_c$, the reduced pressure $P_r = P/P_c$, the reduced density $\rho_r = \rho/\rho_c$, or the reduced molar volume $V_r = V/V_c$. In the vicinity of the critical point there are some general relationships that depend on the deviation of the temperature from the critical point: $\tau = 1 - T_r$. As the critical point is approached from below ($\tau > 0$) the enthalpy of vaporization tends to zero and the heat capacity of the system tends to infinity according to the proportionality relationship

$$C_{\text{P or V}} \propto \tau^{-\alpha} \tag{1.3}$$

The dependence of the density difference between the liquid and vapor phases on τ is according to Eq. (1.4):

$$\Delta\rho = \rho_{\text{liq}} - \rho_{\text{vap}} \propto \tau^{\beta} \tag{1.4}$$

The compressibility of either phase depends on the temperature as

$$d\rho/dP \propto \tau^{-\gamma} \tag{1.5}$$

At $\tau = 0$ (at the critical point) the reduced pressure depends on the reduced density as

$$P_r - 1 \propto (\rho_r - 1)^\delta \tag{1.6}$$

These proportionalities are described by means of the critical indices, the commonly accepted theoretical values for them being $\alpha \approx 0$, $\beta = \frac{1}{2}$, $\gamma = 1$, and $\delta = 3$, assumed to be universal. However, for specific substances the values of these indices differ from the universal ones, as is the case for water (Section 1.5).

The densities of the liquid and vapor phases of a substance approach each other as the critical point is approached. The mean specific volume along the saturation curve is a linear function of the temperature, a nearly perfect experimental fact called the "law of rectilinear diameters":

$$[v_\sigma(l) + v_\sigma(g)]/2 = a + bT \tag{1.7}$$

The value of the temperature coefficient b is generally small and negative, the a and b values for water are presented in Section 1.5.

1.3 SUPERCRITICAL FLUIDS AS SOLVENTS

Supercritical fluids have been proposed as solvents for many uses, both in the laboratory and industrially. Their properties as solvents are, therefore, of interest. The following comparison with gases and liquids (Table 1.2) are illuminating in this respect.

Supercritical fluids have an advantage over ordinary gaseous fluids for many applications in having much higher densities, and an advantage over ordinary liquids in having lower viscosities and considerable higher diffusivities. SCFs are well integrated in the modern tendency toward "green"

TABLE 1.2 Comparison of the Property Ranges of Supercritical Fluids with those of Gases and Liquids at Ambient Conditions

State	Density, $\rho/\mathrm{kg\,m^{-3}}$	Viscosity, $\eta/\mathrm{mPa\,s}$	Diffusivity, $10^6 D/\mathrm{m^2\,s^{-1}}$
Gases, ambient	0.6–2	0.01–0.03	10–40
SCFs, at T_c, P_c	200–500	0.01–0.03	∼0.1
SCFs, at T_c, $4P_c$	400–900	0.03–0.09	∼0.02
Liquids, ambient	500–1600	0.2–3	0.0002–0.002

TABLE 1.3   The Critical Points of Some Substances, Together with their Critical Densities, ρ_c

Substance	T_c/K	P_c/MPa	$\rho_c/kg\,m^{-3}$
Water, H_2O	647.096	22.064	322
Heavy water, D_2O	643.89	21.671	356
Carbon dioxide, CO_2	304.1	7.34	469
Methane, CH_4	190.4	4.60	162
Ammonia, NH_3	405.5	11.35	235
Ethene, C_2H_4	282.4	5.04	215
Ethane, C_2H_6	305.4	4.88	203
Methanol, CH_3OH	512.6	8.09	272
Ethanol, C_2H_5OH	513.9	6.14	276

solvents for reactions and separations that are ecologically friendly. They can be employed as reaction media, for extraction, separation, and purification, and for drug formulations, among other uses. In addition to being "green," they are tunable, so that their properties can be varied at will. The solvent power of supercritical fluids can be fine-tuned by adjustment of the temperature and pressure, hence of the density. This gives them some advantage over common solvents used at ambient conditions, although the solvent power of the latter can be tuned by mixing with cosolvents, as can also be done for SCFs, of course.

Table 1.3 reports the critical temperature, T_c, the critical pressure, P_c, and the critical density, ρ_c, of substances of interest in the present context.

The general mode of the application of SCFs is by dissolution of the reactants or the materials to be separated in them, carrying out the reaction and separation, and then recovering the products by either one of two techniques. One is by rapid diminution of the pressure, allowing the rapid expansion of the supercritical solvent (RESS technique) and eventual formation of the gaseous solvent (for recovery) leaving liquid or solid products behind. The other is the addition of an "antisolvent" that diminishes the solubility of the product in the SCF.

Following are some examples of the uses of SCFs as solvents.

- *Synthesis*: Diels–Alders reactions in supercritical water; polymerization of methyl methacrylate in supercritical difluoromethane; phase transfer catalysis by tetraheptylammonium bromide in supercritical carbon dioxide (SCD) modified with acetone.
- *Pyrolysis*: Conversion of biomass to fuel; complete oxidation of obnoxious materials in SCW (Chapter 5).
- *Separations*: Recovery of fat-soluble vitamins by SCD; delipidation of protein extracts; as chromatographic eluents in forensic analysis; fractionation of mono-, di- and triglycerides.

- *Small Particle Formulations*: Drug particle design, for example, for inhalable insulin; powder drying; microencapsulation.

An important issue for the use of SCFs is their solvent power. One way to describe this is in terms of the linear solvation energy relationship for a series of solutes in a given solvent or of a given solute in a series of solvents:

$$\log x_{\text{solute}} = p + q\rho_{\text{r}} + s\pi^*_{\text{solute}} + a\alpha_{\text{solute}} + b\beta_{\text{solute}} + vV_{\text{solute}} \tag{1.8}$$

Here x_{solute} is the mole fraction of the solute in the saturated solution, V_{solute} is its molar volume, and π^*_{solute}, α_{solute}, and β_{solute} are its Kamlet–Taft solvatochromic parameters, dealing with its polarity/polarizibility, hydrogen bond donation (or electron pair acceptance) ability, and hydrogen bond acceptance (or electron pair donation) ability [2]. The parameters p, q, s, a, b, and v characterize the SCF, the latter four depending on the reduced density, ρ_{r}. The parameter v is generally negative, solute solubilities decrease with increasing solute molar volumes. Some SCFs are apolar (e.g., SCD), that is, they are incapable of acceptance and donation of hydrogen bonds (or electron pairs), hence have $a = b = 0$ and their s values may be near zero or negative. Others are dipolar aprotic (like acetone) with $b = 0$ but have positive values for a and s, and still others are protic (like water and methanol) and have also $b > 0$, being capable of both acceptance and donation of hydrogen bonds.

Another way to describe the solvent power of an SCF is by means of its (Hildebrand) solubility parameter, δ_{H} [3]. For solute/solvent systems with small solute solubilities, the following expression

$$\ln x_{\text{solute}} = -V_{\text{solute}}(RT)^{-1}(\delta_{\text{H SCF}} - \delta_{\text{H solute}})^2 \tag{1.9}$$

is valid. Conformal fluids, that is, SCFs conformal with a reference fluid, generally interact through central forces of the van der Waals type. Their solubility parameter can be expressed by

$$\delta_{\text{H SCF}}/\text{MPa}^{1/2} = 2.15 P_{\text{c}}^{1/2} T_{\text{r}}^{-1/4} \rho_{\text{r}} \tag{1.10}$$

They depend on the temperature and pressure mainly through the dependence of the reduced density, ρ_{r}. The solubility parameters of the solutes are available from group contributions and are assumed to be rather independent of the temperature and pressure for estimation of the solubility via Eq. (1.9). A difference $\delta_{\text{H SCF}} - \delta_{\text{H solute}} > 4\,\text{MPa}^{1/2}$ represents a low solubility of the solute. However, water is *not* a conformal fluid, since it interacts by dipole and hydrogen bonding interactions beyond the van der Waals ones. Its solubility parameter is discussed in Section 2.6 and the resulting expression, Eq. (2.42), differs from Eq. (1.10).

The best-known and most widely employed supercritical fluid is SCD, used industrially on a large scale for the decaffeination of coffee. It has convenient T_c and P_c values: 30.9°C (304.1 K) and 72.4 atm (7.34 MPa), is environmentally friendly (its industrial contribution to the greenhouse gas inventory is negligible compared to that resulting from the burning of fossil fuels), is nonflammable, nontoxic (at low concentrations), and noncorrosive to common structural materials, and is cheap. Its main drawback is its low solvent power, the solubility of polar substances in it is rather small and even nonpolar substances are not very soluble in SCD. This situation can be ameliorated by the addition of modifiers, such as methanol, but a detailed discussion of SCD, its properties and uses, is outside the scope of this book.

Supercritical water has properties differing considerably from those of SCD; for some recent reviews, see Refs [4, 5]. Its critical (absolute) temperature is twice and its critical pressure almost three times those of SCD; its ranges of application are about 400–600°C and 30–100 MPa. Thus, although nontoxic, nonflammable, and environmentally friendly it involves certain technical difficulties in view of these high temperatures and pressures, and mainly also corrosion problems. Applications of SCW and problems involved with them are fully discussed in Chapter 5. Here the solvent powers are only mentioned, these being fully discussed in Chapter 4: SCW retains the hydrogen bond donation and acceptance abilities of the water molecule, and its relatively low permittivity permits dissolution of nonpolar solutes.

1.4 GASEOUS AND LIQUID WATER

Before going on to the properties of supercritical water, it is expedient to survey briefly the properties of water in the gas phase, as individual molecules and clusters, and in the liquid phase, in the latter case along the saturation curve (Fig. 1.1). Some of the properties of individual water molecules [6, 7], both H_2O and D_2O, are shown in Table 1.4. These properties are manifested when the water molecules are present at low pressures in the ideal gaseous state, where $PV/RT = 1$.

The nonideality of water vapor (gaseous water) at nonnegligible pressures can be expressed in terms of the compressibility factor Z and the virial expansion:

$$Z = PV/RT = 1 + B_2(T)V^{-1} + B_3(T)V^{-2} + \cdots \qquad (1.11)$$

TABLE 1.4 Some Molecular Properties of (Light) Water, H_2O and Heavy Water, D_2O

Property	H_2O	D_2O
Molar mass, M/kg mol^{-1}	0.018015	0.020031
O—H(D) bond length, d/pm	95.72	95.75
H—O—H angle, °	104.523	104.474
Moment of inertia,$^a I_A$/10^{-30} kg m^{-2}	0.10220	0.18384
Moment of inertia,$^a I_B$/10^{-30} kg m^{-2}	0.19187	0.38340
Moment of inertia,$^a I_C$/10^{-30} kg m^{-2}	0.29376	0.56698
Hydrogen bond length, d/pm	276.5	276.6
Dipole moment, μ/D^b	1.834	1.84
Electrical quadrupole moment, Θ/10^{-39} C m^2	1.87	
Polarizability, α/10^{-30} m^3	1.456	1.536
Collision diameter, σ/pm	274	
Potential energy minimum, (u/k_B)/K	732	
O—H(D) bond energy at 0 K, E/kJ mol^{-1}	44.77	
Proton (deuteron) affinity, E/kJ mol^{-1}	762^c	772^c
Symmetrical stretching frequency, v_1/cm^{-1}	3656.65	2671.46
Asymmetrical stretching frequency, v_3/cm^{-1}	3755.79	2788.05
Bending frequency, v_2/cm^{-1}	1594.59	1178.33
Zero point vibrational energy, kJ mol^{-1}	55.31^d	40.44^d
Ideal gas heat capacity, C_p^{ig}/J · K^{-1} · mol^{-1}	33.584	
Ideal gas entropy, S^{ig}/J K^{-1} mol^{-1}	188.67	

^{a}The moment of inertia I_A pertains to rotation round the O—H bond and I_B pertains to rotation round the axis through the oxygen atom, and I_C pertains to rotation round an axis perpendicular to the latter.
^{b}1 D = 3.33564×10^{-30} C m.
^{c}From Ref. [45].
^{d}From Ref. [46].

The temperature-dependent B_i are called the second, third, and so on, virial coefficients. The experimental B_2 values from 50 to 460°C are expressed by [7]

$$B_2/\text{cm}^3 \text{mol}^{-1} = 37.15 - (5.1274 \times 10^4/(T/\text{K}))\exp(1.7095 \times 10^5/(T/\text{K})^2)$$

(1.12)

It is negative ($-1059 \text{ cm}^3 \text{mol}^{-1}$ at 20°C) but becomes less so as the temperature increases ($-292 \text{ cm}^3 \text{mol}^{-1}$ at 150°C and $-105 \text{ cm}^3 \text{mol}^{-1}$ at 325°C). Values of the third virial coefficient of water are less well known, B_3 is zero at < 150°C, it is about $-5.9 \times 10^4 \text{ cm}^6 \text{mol}^{-1}$ at 150°C and changes sign near 300°C [8].

Water in the gaseous phase, unless at very low pressures, tends to associate to dimers, trimers, and larger clusters. The clustering to dimers and higher oligomers can be regarded as an interpretation of the nonideality of the water vapor expressed by Eq. (1.11). Rowlinson [9] derived the equilibrium constant for the association of water molecules to dimers in terms of the partial pressures of the species as

$$\log(K_{ass}/Pa) = -3.644 + 1250/(T/K) \qquad (1.13)$$

This corresponds to a (temperature independent) enthalpy of dimerization of $-23.9\,kJ\,mol^{-1}$ and would lead to the presence of a mole fraction of 0.04 for the dimers at the critical point [7]. Less negative enthalpies of dimerization were consistent with the speed of ultrasound, $-15.4 \pm 8.7\,kJ\,mol^{-1}$ [10] and values varying from $-16.3\,kJ\,mol^{-1}$ at 0°C down to $-11.4\,kJ\,mol^{-1}$ at the critical point were derived from computer simulations by Slanina [11]. However, a more recent estimate by Sit et al. [12] of the binding energy of the dimer is $-22.8\,kJ\,mol^{-1}$. A consequence of the uncertainty in the dimerization enthalpy is an uncertainty in the mole fraction of the dimer as pressure and density of the vapor (and the temperature) increase along the saturation line. The increased total pressure should enhance the dimerization, and mole fractions of $x_{dim} = 0.276$ and $x_{trim} = 0.006$ have been estimated by Slanina [13] for the dimer and trimer at the critical point.

The pressure dependence of the integrated infrared absorption intensities of the stretching modes $v_1 + v_3$ of water vapor (3200–4200 cm^{-1}) at 300–450°C and up to 8 MPa is consistent with a low content of dimers. The resulting dimerization enthalpy $-16.7 \pm 3.8\,kJ\,mol^{-1}$ [14] is within the range of the values quoted above.

The Raman spectra of saturated water vapor were measured by Walrafen et al. [15], showing an isosbestic point at 3648.5 cm^{-1} as the temperature is increased. Frequencies above this point are considered due to monomeric water, those below it, their intensities being proportional to the square of the density, are considered due to the dimers. A much higher mole fraction of dimers at the critical point results from the Raman data, namely $x_{dim} = 0.696$. This interpretation was subsequently withdrawn by the authors [16], since the resulting enthalpies of dimerization had the wrong sign. The two-species equilibrium suggested by the isosbestic point was attributed to low and high rotationally excited monomeric water molecules. This does not explain the proportionality of the low frequency Raman intensities to the square of the pressure (density), however.

Dimers are not considered to contribute appreciably to the infrared absorption of dilute water vapor in the atmosphere, whereas large clusters of, say, 30 molecules do [17]. These are of consequence to cloud formation and consist of multihydrated hydrogen and hydroxide ions: $H(H_2O)_n{}^+$ and $HO(H_2O)_m{}^-$, or even the neutral ion pairs or zwitterions $H^+(H_2O)_pOH^-$.

Clusters of water around other ions are found by mass spectrometry. A further consideration of these entities, important for cloud formation, is outside the scope of this book.

Turning now to liquid water along the saturation line, some of the relevant properties are collected in Table 1.5.

Several features of the property changes of the liquid water with increasing temperatures and diminishing densities, ρ, along the saturation line should be noted. The relative permittivity, ε_r, the dynamic viscosity, η, the surface tension, γ, and the molar enthalpy of vaporization, $\Delta_V H$, diminish steadily as the temperature is increased. On approaching the critical point, say above 300°C, they fall rather abruptly, the latter two quantities, γ and $\Delta_V H$ vanishing at the critical point. On the other hand, the constant pressure molar heat capacity, C_P, and the isothermal compressibility, κ_T, have a shallow minimum near 33 and 46°C, respectively, but increase with the temperature, diverging to infinity at the critical point.

Several derived quantities are of interest; for instance, the temperature and pressure derivatives of the relative permittivity have been reported by Fernandez et al. [18]. These yield according to Bradley and Pitzer [19] the limiting slopes for the Debye–Hückel theoretical expressions for the osmotic and activity coefficients and the apparent molar enthalpies and volumes of electrolytes in water up to 350°C along the saturation line as well at higher pressures. The complex permittivity of water was measured by Lyubimov and Nabokov [20, 21] up to 260°C along the saturation curve. The derived relaxation times decrease steadily from 5.82 ps at 40°C to 0.75 ps at 260°C. The corresponding values for heavy water, D_2O, are 25% to 12% larger.

Other derived quantities include the isobaric expansibility, α_P, or the pressure derivative of the compressibility itself (the second pressure derivative of the density), the isochoric (constant volume) molar heat capacity, C_V, and the speed of sound, u. These are to be found in the Steam Tables and they can be derived from the equation of state as presented most recently by Wagner and Pruss [1]. The internal pressure, $P_i = T\alpha_P/\kappa_T - p_\sigma$, has a maximum of 758 MPa near 172°C, diminishing to 473 MPa at 360°C. The cohesive energy density $(\Delta_V H° - RT)/V$ decreases steadily from 2357 MPa at 0°C to 1653 MPa at 172°C to 236 MPa at 360°C. A liquid is deemed structured ("stiff") when the cohesive energy density is larger by ≥ 50 MPa than the internal pressure, as suggested by Marcus [22]. Since there is a crossover between these two dependencies by 328°C, liquid water above this temperature is no longer "stiff" according to this criterion.

An interesting derived quantity is the Kirkwood–Fröhlich dipole orientation parameter g, which reflects the association of the water molecules in the liquid. For a coordination number N_{co} (the number of nearest neighbors), obtainable experimentally from diffraction of X-rays or neutrons, g describes

TABLE 1.5 Some Properties of Liquid Water Along the Saturation Line

$t/^\circ\mathrm{C}$	p_σ/MPa	$\rho_\sigma/\mathrm{kg\,m^{-3}}$	ε_r	$\eta/\mathrm{mPa\,s}$	$\Delta_\mathrm{V}H/\mathrm{kJ\,mol^{-1}}$	$\gamma/\mathrm{mN\,m^{-1}}$	$C_\mathrm{P}/\mathrm{J\,K^{-1}\,mol^{-1}}$	$\kappa_\mathrm{T}/\mathrm{GPa^{-1}}$	g	$\mathrm{p}K_\mathrm{W}$[a]
25	0.00317	997.05	78.46	0.890	44.04	71.96	75.29	0.4525	2.90	14.00
50	0.01235	988.00	69.90	0.547	43.09	67.93	75.31	0.4417	2.84	13.28
75	0.03856	974.82	62.24	0.374	42.24	63.54	75.53	0.4561	2.77	12.71
100	0.10133	958.36	55.41	0.282	41.50	58.78	75.95	0.4902	2.70	12.26
125	0.23203	939.05	49.36	0.218	39.2	53.79	76.70	0.549[b]	2.64	11.91
150	0.47574	917.04	43.94	0.182	38.1	48.70	77.63	0.634[b]	2.58	11.64
175	0.8926	894.5	39.16	0.143	36.7	43.28	78.92	0.745[b,c]	2.52	11.44
200	1.5551	865.86	34.74	0.134	35.0	37.81	80.75	0.883[b,c]	2.48	11.30
225	2.5505	834.3	30.80	0.118	32.8	32.03	83.44		2.45	11.22
250	3.9777	799.14	27.08	0.1062	30.5	36.18	87.36		2.38	11.20
275	5.949	741.0	23.63	0.0967	28.1	20.17	92.98		2.33	11.25
300	8.592	712.42	20.26	0.0859	25.3	14.39	100.87		2.25	11.38
325	12.058	649.0	15.84	0.0820	22.2	7.26	111.66		2.15	11.64
350	16.537	575.04	13.16	0.0659	16.0	3.79	126.10		2.00	12.14
373.95	22.064	317.76	5.44		0	0	∞		1.79	15.34[d]

Adapted from Refs [20, 49].
[a]From Ref. [23].
[b]From Ref. [47].
[c]Extrapolated with a quadratic fit to data at 0–150°C.
[d]From Ref. [48].

the average angle, θ_{ij}, between the dipoles of neighboring water molecules i and j:

$$g = 1 + N_{co}\langle\cos\theta_{ij}\rangle \tag{1.14}$$

The values of g are obtained from the measured relative permittivity according to

$$g = (9\varepsilon_0 k_B/N_A)(MT/\rho\mu^2)(\varepsilon_r - \varepsilon_\infty)(2\varepsilon_r + \varepsilon_\infty)/\varepsilon_r(\varepsilon_\infty + 2)^2 \tag{1.15}$$

Here ε_0, k_B, and N_A are the universal constants, M is the molar mass of water, ρ is its density (M/ρ is its molar volume, V), μ is its permanent dipole moment (Table 1.4), and ε_∞ is the infinite frequency permittivity, equal to the square of the infinite frequency refractive index, n_∞^2. The latter quantity is obtained from the polarizability of water, α (Table 1.4) according to the Lorenz–Lorentz relation as

$$n_\infty^2 = (V + 2B\alpha)/(V - B\alpha) \tag{1.16}$$

where $B = 4\pi N_A/3$. The values of g, shown in Table 1.5, are calculated according to the polynomial reported by Marcus [22]. The g values > 1 show that water is associated as a liquid, though to a diminishing extent as the temperature rises. The average angle, with an assumed (not much temperature dependent) coordination number $N_{co} = 4$, increases from 61.6° at 25°C to 78.3° at 200°C and 78.6° at the critical point, showing some alignment of the dipoles even at these high temperatures (an angle of 90° corresponds to no correlation between the dipoles and to no association).

The structuredness of water has been explored by the present author [22] in terms of several measures. One of these is the ratio of the difference of the standard molar Gibbs energies of condensation of light and heavy water to the difference in the hydrogen bond energies of these two kinds of water. The resulting average number of hydrogen bonds per water molecule diminishes from 1.34 at 5°C to 0.30 at 305°C, but there are several difficulties with this interpretation of the structuredness, as discussed in the cited paper. The heat capacity density of liquid water relative to its ideal gas is another measure of the structuredness. It changes only slightly, from 2.37 to 2.20 J/K^{-1} cm^{-3} from 0 to 200°C, and remains considerably higher than for unstructured liquids, for which it is $\sim$0.6 J/K^{-1} cm^{-3}. The entropy deficit of liquid water relative to its vapor (corrected for association of the latter) and normalized with regards to this quantity for the unstructured methane, $\Delta\Delta_V S^\circ/R$, changes more from 8.35 at 0°C to 6.93 at 200°C and to 6.06 at 360°C, again being much larger than for unstructured liquids, for which it is 2.0. That is, water is "ordered" according to this criterion. This measure of structuredness is quite well proportional to the dipole orientation parameter through the relevant temperature range: $(\Delta\Delta_V S^\circ/R)/g = 2.83 \pm 0.15$, both measures pertaining to the "order" in the liquid water [22].

Another quantity of interest is the ionic dissociation constant of the water, or rather the ion product $K_W = [H_3O^+][OH^-]$. The recent evaluation of Chen et al. [23] let to the values of pK_W (for K_W on the $(mol\ kg^{-1})^2$ scale) shown in Table 1.5. This reference provides also values for the enthalpy, entropy, heat capacity and volume changes for the ionization of water at its saturation pressure. Note that pK_W decreases (ionization increases) as the temperature increases up to $\sim 250^\circ C$ and increases again (ionization decreases) somewhat beyond this temperature.

Spectroscopic studies of liquid water along the saturation line reveal some interesting aspects. The overtone band of HOD shows features at 7000 and $7150\ cm^{-1}$, attributed by Luck [24] to nonhydrogen bonded water molecules, a band at $6850\ cm^{-1}$ attributed to weakly cooperative hydrogen bonded molecules, and one at $6400\ cm^{-1}$ attributed to strongly cooperative hydrogen bonded molecules. As expected, the latter diminish sharply from 52% at $0^\circ C$ to 7% at $230^\circ C$ and remain near that level up to $350^\circ C$. On the other hand, the nonhydrogen bonded molecules increase from $\sim 10\%$ at $0^\circ C$ rather steadily to $\sim 65\%$ at $350^\circ C$. The ^{17}O NMR chemical shifts δ were measured by Tsukahara et al. [25] in pressurized water (30 MPa) from $25^\circ C$ up to and beyond the critical point. The values of δ/ppm vary substantially linearly from -1 at $25^\circ C$ to -18 at $350^\circ C$ and the deduced extent of hydrogen bonding, relative to that in ambient water, η, is linear with it: $\eta = 1 + 0.0274(\delta/ppm)$, thus diminishing to 0.51 at $350^\circ C$. The deduced reorientation relaxation times diminish sharply from 2.9 ps at $25^\circ C$ to 0.6 ps at $100^\circ C$ and then more gently to 0.1 ps at $350^\circ C$. These values may be compared with the dielectric relaxation times reported above.

Thermodynamic properties of heavy water, D_2O, along its saturation line have been reviewed and compared to those of light (ordinary) water, H_2O, by Hill and MacMillan [8] up to $325^\circ C$. The data shown are the liquid and vapor volumes (densities) and enthalpies and the specific heat of the liquid. Note that the critical point of D_2O is lower than that of H_2O by 3.21 K (Table 1.3).

As mentioned in Section 1.3, solvatochromic parameters can be used to predict the solubility of solutes in solvents according to Eq. (1.8). Lu et al. [26] reported plots of the Kamlet–Taft π^*, α and β solvatochromic parameters as well as those of Reichardt's $E_T(30)$ polarity index in water along the saturation line up to $275^\circ C$. One of these parameters, π^* that describes the polarity/polarizibility of the solvent, measured by means of the solvatochromic indicator 4-nitroanisole, is linear with the density of the water. Its temperature dependence curve is concave downward (π^* diminishing from 1.09 at ambience to 0.69 at 275°C), whereas that of the hydrogen bond donation parameter α is concave upward (diminishing from 1.19 at ambience to 0.84 at 275°C). The hydrogen bond acceptance parameter β (for monomeric water) varies little up to 100°C, then increases by some 30% up to 150°C and remains more or less at this value up to 275°C. The polarity index $E_T(30)$ decreases linearly from 63 kcal mol^{-1} (264 kJ mol^{-1}) at ambience to 53 (222 kJ mol^{-1}) at 275°C. In pressurized (40 MPa) high temperature water [27] (up to 420°C, i.e., beyond the critical point), the values of π^* were again linear with the density of the water, $\pi^* = 1.77\rho - 0.71$ up to 360°C. If the pressure dependence of π^* for the liquid is assumed to be negligible, the result is $\pi^* = 0.980 + 7.12 \times 10^{-4}(t/°C) - 7.40 \times 10^{-6}(t/°C)^2$ for water along the saturation line. The alternative probes, 4-nitro- and 4-cyano-N,N-dimethylaniline yield values consistent with these for liquid water.

1.5 NEAR-CRITICAL WATER

There is no clear definition of "near-critical" water, but most publications tend to use this epithet for pressurized hot liquid water at temperatures $300 \leq t/°C \leq 375$. Near-critical water has been proposed as an environmentally benign medium for chemical reactions [28, 29], biomass decomposition [30] and gasification [31], or preparation of inorganic nanoparticles [32], to cite at random some recent applications.

As the critical point is approached the differences in properties between the liquid and the vapor diminish, the fluid becomes opalescent due to the minute dispersion of liquid-like and vapor-like domains, and the meniscus separating the liquid from the vapor vanishes at the critical point. There remaining no free surface at the critical point, the surface tension γ vanishes, and since no energy is needed to convert the liquid to its vapor (these states being identical) also the enthalpy of vaporization, $\Delta_V H$, vanishes and the heat capacities C_P and C_V diverge to infinity (Table 1.5). Still, the properties of water near the critical point, say within 1–5 K above and below it, have been studied intensively.

The law of rectilinear diameters, Eq. (1.7) applies to water as for other fluids. For water the values of the coefficients of the mean specific volumes are $a = (61.6 \pm 0.6) \times 10^{-3}$ m^3 kg^{-1} and $b = (90.4 \pm 1.0) \times 10^{-6}$ m^3/kg$^{-1} \cdot$ K^{-1} over a considerable range of temperatures: $T \geq 600$ K up to the critical point. The critical density derived from this expression is the accepted value of 322 ± 3 kg m^{-3}.

The critical indexes (see Section 1.2) are supposed to have universal values, but when actual P, ρ, and T (PVT) data for water very near the critical point, at $\tau = 1 - T/T_c \sim 0.01$ are used somewhat different values result: $\alpha = 0.11$ (instead of ≈ 0), $\beta = 0.34$ (instead of 0.5), $\gamma = 1.21$ (instead of 1), and

$\delta = 4.56$ (instead of 3) [33]. A slightly different set was given earlier by Watson et al. [34]: $\alpha = 0.087$, $\beta = 0.335$, $\gamma = 1.212$, and $\delta = 4.46$. With these and 13 additional parameters the equation of state is established from $\tau = -0.002$ to 0.118 within the experimental errors of the PVT data. Also, the sound velocity u and heat capacity C_P data could be modeled well around the critical density. Considerably fewer parameters (only eight) are needed when the PVT data are modeled according to the three-site associated perturbed anisotropic chain theory (APACT), with a total mean deviation of $\sim 1\%$ for the density and $\sim 2\%$ for the pressure for $-0.27 \leq \tau \leq 1.97$ [35].

Supplementary values for $V(T, P)$ as well as the enthalpy and entropy of water at the near-critical region were reported by Kretzschmar et al. [36, 37], based on the industrial formulation of the IAPWS-97. These are of main interest for steam turbine calculations and need not be dealt with here.

The molar isochoric heat capacity (that at constant volume), C_V, was reported by Abdulagatov et al. [50] over the temperature range 140–420°C and at densities of 250–925 kg m^{-3}, encompassing the critical point, at which it diverges to infinity. More recently Abdulagatov et al. [51] drew the attention to the extrema of C_V. At the critical density of 322 kg m^{-3} and $\tau = -0.055$, $C_V = 63$ J/K^{-1} mol^{-1} and this value is reached again beyond the critical point at $\tau = 0.042$. In the two-phase region and at densities near the critical one C_V/J/K^{-1} mol^{-1} increases from 63 at $\tau = -0.055$ to 193 at $\tau = -0.039$ J/K^{-1} mol^{-1} to 234 J/K^{-1} mol^{-1} at $\tau = -0.0093$ [50].

The thermal conductivity of water shows a critical enhancement near the critical point [38] as does its viscosity, as discussed by Watson et al. [34]. The viscosity enhancement takes place in a region bounded by $|\tau| \leq 0.023$ and $|\rho_r - 1| \leq 0.25$ and a 22-parameter expression describes the viscosity of water both outside this interval (η'), and inside it, $\eta = \eta' + \Delta\eta$, including the enhancement, $\Delta\eta$. This depends on the density fluctuations near the critical point, described by the correlation length ξ, so that

$$\Delta\eta = \eta'(0.376\xi)^{0.054} - 1 \qquad (1.17)$$

with ξ in nm. At 373.5°C and saturation pressure, the viscosity of liquid water is 47.02 μPa s and at 375°C and at 22.5 MPa (slightly above the critical point temperature and pressure, $t_c = 373.95$°C and $P_c = 22.06$ MPa) it is 41.79 μPa s, whereas in between it diverges to infinity [34].

The permittivity of water does not show a discontinuity near the critical point, it varies smoothly along the saturation curve and beyond the critical point [18]. On the other hand, the Raman scattering of the ν_1 stretching vibration of water shows according to Ikushima et al. [39] a moderate jump in the blue shift with increasing temperatures in the near-critical region as well a strong narrowing of the bandwidth. These changes are consistent with the changes in the proton chemical shifts of the NMR spectrum measured by Matsubayashi et al. [40] that also show such jumps. The results were interpreted in a breakdown of the hydrogen bonded oligomeric structures in the near-critical region, leaving only hydrogen bonded dimers in addition to the monomers.

Turning now to heavy water, D_2O, and its mixtures with ordinary water, H_2O, the composition dependence of the critical temperature was measured by Marshall and Simonson [41] and is linear with the mole fraction of the former:

$$T_c/K = 647.14 - 3.25x_{D_2O} \tag{1.18}$$

The critical molar volumes of the two pure components, and presumably also of the HDO produced in the mixtures

$$D_2O + H_2O \leftrightarrows 2HDO \tag{1.19}$$

are the same, $56.0 \pm 0.1 \, cm^3 \, mol^{-1}$. A crossover equation of state (see Section 2.1.3) was presented by Kiselev et al. [42] for D_2O and its mixtures with H_2O that fitted the data well also in the near-critical region. In a subsequent paper Bazaev et al. [43] refined the PVT data and Polikhronidi et al. [44] did the same for the isochoric heat capacity, C_V, for the equimolar mixture $D_2O + H_2O$ in the near-critical region.

1.6 SUMMARY

The phase diagram of a single pure substance reports the external conditions (temperature and pressure) at which two (or at most three) phases are at equilibrium with each other. The phase diagram of water is shown in Fig. 1.1. The vapor–liquid equilibrium line, showing the vapor pressure p_σ of water along the saturation curve is shown in Fig. 1.1 too, and is parametrized as Eq. (1.2) with parameters shown in Table 1.1. A supercritical fluid consists of a single phase and has two degrees of freedom, the temperature and the pressure can be chosen at will, provided they are larger than the critical values.

The critical point of fluids in general is discussed, the critical temperature of water being $T_c = 647.096 \, K = 373.946°C$ and its critical pressure is $P_c = 22.064 \, MPa = 21.78 \, atm$. The critical density of water is $\rho_c = 322 \, kg \, m^{-3}$ and its critical molar volume is $V_c = 56.0 \times 10^{-6} \, m^3 \, mol^{-1}$. The thermodynamic states of the supercritical fluids are often expressed in terms of the reduced quantities, that is, $T_r = T/T_c$, and so on. In the vicinity of the critical point there are some general relationships that depend on the deviation of the temperature from the critical point: $\tau = 1 - T_r$. The thermodynamic functions, such as the heat capacity and compressibility, are then expressed in terms of the critical indices that are exponents of τ. The mean specific volume (of liquid + vapor) along the saturation curve is a linear function of the temperature, the "law of rectilinear diameters," Eq. (1.7), that extrapolates to the critical density.

Supercritical fluids have valuable properties that make some of them "green," that is, environmentally friendly solvents. An important property is their being tunable, the freely disposable pressure and temperature determine the density. Critical points of some useful SCFs are shown in Table 1.3 and the advantages of SCFs relative to liquids and gases are shown in Table 1.2. The solvent power of SCFs can be expressed in terms of the linear solvation energy relationship (1.8) or in terms of the solubility parameter expression (1.9). Some applications of SCFs are listed, but those of SCW are fully discussed in Chapter 5.

The properties of water below the critical point are dealt with for water vapor (gaseous water) and liquid water. At non-negligible pressures the pressure–volume–temperature (*PVT*) dependence of water vapor can be expressed in terms of the compressibility factor Z and the virial expansion (1.11). Clustering of water vapor molecules to dimers and higher oligomers can be regarded as an interpretation of the nonideality of the water vapor, Eq. (1.11). Uncertainty of the enthalpy of dimerization, as derived by several authors using different methods, for example, the partial pressure and the speed of ultrasound, leads to a great uncertainty of the extent of dimerization of the water vapor. The pressure dependence of the integrated infrared absorption intensities of the O–H bond stretching is consistent with a low content of dimers, but an interpretation of the corresponding dependence of the Raman scattering intensity yields an apparent large dimer content.

Some of the relevant properties of liquid water along the saturation line are collected in Table 1.5. The relative permittivity, ε_r, the dynamic viscosity, η, the surface tension, γ, and the molar enthalpy of vaporization, $\Delta_V H$, diminish steadily as the temperature is increased. On approaching the critical point, say above 300°C, they fall rather abruptly, the latter two quantities, γ and $\Delta_V H$ vanishing at the critical point, but the constant pressure molar heat capacity, C_P, and the isothermal compressibility, κ_T, increase with the temperature, diverging to infinity at the critical point. The ion product of water, $K_W = [H_3O^+][OH^-]$, describing its ionic dissociation along the saturation line, is shown in Table 1.5 as a function of the temperature.

Liquid water is deemed structured ("stiff") since its cohesive energy density is larger by ≥ 50 MPa than its internal pressure. However, there is a crossover between these two dependencies by 328°C, so that liquid water above this temperature is no longer "stiff" according to this criterion. The Kirkwood dipole orientation parameters g, Eq. (1.15), are shown in Table 1.5. The g values >1 show that water is associated as a liquid, though to a diminishing extent as the temperature rises. The heat capacity density of liquid water relative to its ideal gas is another measure of the structuredness, being considerably higher than for unstructured liquids. The entropy deficit of liquid water relative to its vapor (corrected for association of the latter) and normalized with regards to this quantity for the unstructured methane is also much larger than for unstructured liquids, so that water is "ordered" according to these criteria. The extent of hydrogen bonding in liquid water can be derived from the difference in the Gibbs energies of condensation of light and heavy water and from the NMR chemical shifts of water. Both measures show a steady decrease as the temperature is raised, but this subject is more fully discussed in Section 3.4.

There is no clear definition of "near-critical" water, but most publications refer so to pressurized hot liquid water at temperatures $300 \leq t/^\circ C \leq 375$. Near-critical water has been proposed as an environmentally benign ("green") medium. The solvatochromic parameters of water, relevant to Eq. (1.8), have been determined to 275°C, that is, still short of near-critical water. Values of the coefficients of the "law of rectilinear diameters," Eq. (1.7) and of the critical indices (exponents of τ, the distance from T_c) of near-critical water are presented in the text. The thermal conductivity and viscosity of water exhibit an enhancement in near-critical water, and detailed considerations of its isochoric heat capacity are also dealt with.

REFERENCES

1. W. Wagner and A. Pruss, *J. Phys. Chem. Ref. Data* **31**, 387 (2002).
2. S. Yalkowsky, *Aqueous Solubility: Methods of Estimation for Organic Compounds*, Marcel Dekker, New York, 1992.
3. Y. Marcus, *J. Supercrit. Fluids* **38**, 7 (2006).
4. M.-C. Bellisent-Funel, *J. Mol. Liq.* **90**, 313 (2001).
5. H. Weingärtner, *Angew. Chem. Intl. Ed.* **44**, 2673 (2005).
6. Y. Marcus, *The Properties of Solvents*, Wiley, Chichester, 1998.
7. D. Eisenberg and W. Kauzmann, *The Structure and Properties of Water*, Clarendon Press, Oxford, 1969.
8. P. G. Hill and R. D. C. Whalley, *J. Phys. Chem. Ref. Data* **9**, 735 (1980).
9. J. S. Rowlinson, *Trans. Faraday Soc.* **47**, 974 (1949).
10. R. A. Bolander and H. A. Gebbie, *Nature* **253**, 523 (1975).
11. Z. Slanina, *J. Mol. Struct.* **273**, 81 (1990).
12. P. L.-L. Sit and N. Marzari, *J. Chem. Phys.* **122**, 204510 (2005).
13. Z. Slanina, *Thermochim. Acta* **116**, 161 (1987).
14. G. V. Bondarenko and Yu. E. Gorbaty, *Mol. Phys.* **74**, 639 (1991).

15. G. E. Walrafen, W.-H. Yang, and Y. C. Chu, *J. Phys. Chem. B* **105**, 7155 (2001).

16. G. E. Walrafen, W.-H. Yang, and Y. C. Chu, *J. Phys. Chem. B* **103**, 1332 (2001).

17. H. R. Carlon, *J. Appl. Phys.* **52**, 3111 (1981).

18. D. P. Fernandez, A. R. H. Goodwin, E. W. Lemmon, J. M. H. Levelt Sengers, and R. C. Williams, *J. Phys. Chem. Ref. Data* **26**, 1125 (1997).

19. D. J. Bradley and K. S. Pitzer, *J. Phys. Chem.* **83**, 1599 (1979).

20. Yu. A. Lyubimov and O. A. Nabokov, *Russ. J. Phys. Chem.* **59**, 86 (1985).

21. O. A. Nabokov and Yu. A. Lyubimov, *Russ. J. Phys. Chem.* **61**, 106 (1987).

22. Y. Marcus, *J. Mol. Liquids* **79**, 151 (1999).

23. X. Chen, J. L. Oscarson, S. E. Gillespie, H. Cao, and R. M. Izatt, *J. Solution Chem.* **23**, 747 (1994).

24. W. A. P. Luck, *J. Mol. Struct.* **448**, 131 (1998).

25. T. Tsukahara, M. Harada, H. Tomiyasu, and Y. Ikeda, *J. Supercrit. Fluids* **26**, 73 (2003).

26. J. Lu, J. S. Brown, E. C. Boughner, C. L. Liotta, and C. A. Eckert, *Ind. Eng. Chem. Res.* **41**, 2835 (2002).

27. K. Minami, M. Mizuta, M. Suzuki, T. Aizawa, and K. Arai, *Phys. Chem. Chem. Phys.* **8**, 2257 (2006).

28. H. R. Patrick, K. Griffith, C. L. Liotta, C. A. Eckert, and R. Glaeser, *Ind. Eng. Chem. Res.* **40**, 6063 (2001).

29. Z. Dai, B. Hatano, and H. Tagaya, *Appl. Catal. A. Gen.* **258**, 189 (2004).

30. A. Sinag, S. Gülbay, B. Uskan, and M. Canel, *Energy Conv. Manag.* **51**, 612 (2010).

31. P. Azadi, K. M. Syed, and R. Farnood, *Appl. Catal. A. Gen.* **358**, 65 (2009).

32. P. Hald, M. Bremholm, S. B. Iversen, and B. B. Iversen, *J. Solid State Chem.* **181**, 2681 (2008).

33. M. A. Anisimov, S. B. Kiselev, and I. G. Kostyukova, *Teplof. Vysok. Temp.* **25**, 31 (1987).

34. J. T. R. Watson, R. S. Basu, and J. V. Sengers, *J. Phys. Chem. Ref. Data* **9**, 1255 (1980).

35. P. J. Smits, I. G. Economou, C. J. Peters, and J. de S. Arons, *J. Phys. Chem.* **98**, 12080 (1994).

36. H.-J. Kretzschmar, I. Stocker, T. Willkommen, J. Trubenach, and A. Dittmann, in P. R. Tremaine, Ed., *Steam, Water, and Hydrothermal Systems: Physics and Chemistry Meeting the Needs of Industry*, National Research Council Canada, Ottawa, 2000, p. 255.

37. H.-J. Kretzschmar, J. R. Cooper, J. S. Gallagher, A. H. Harvey, K. Knobloch, R. Mares, K. Miyagawa, N. Okita, R. Span, I. Stocker, W. Wagner, and I. Weber, *J. Eng. Gas Turb. Power* **129**, 294, 1125 (2007).

38. R. S. Basu and J. V. Sengers, in A. Cezairlian, Ed., *Proceedings of the 7th Symposium on Thermophysical Properties*, American Society of Mechanical Engineers, New York, 1977, p. 822.

39. Y. Ikushima, K. Hatakeda, N. Saito, and M. Arai, *J. Chem. Phys.* **108**, 5855 (1998).

40. N. Matsubayashi, C. Wakai, and M. Nakahara, *J. Chem. Phys.* **107**, 9133 (1997).

41. W. L. Marshall and J. M. Simonson, *J. Chem. Thermodyn.* **23**, 613 (1991).

42. S. B. Kiselev, A. M. Abdulagatov, and A. H. Harvey, *Int. J. Thermophys.* **20**, 563 (1999).

43. A. R. Bazaev, I. M. Abdulagatov, J. W. Magee, E. A. Bazaev, and A. E. Ramazanova, *J. Supercrit. Fluids* **26**, 115 (2003).

44. N. G. Polykhronidi, I. M. Abdulagatov, J. W. Magee, and G. V. Stepanov, *Int. J. Thermophys.* **24**, 405 (2003).

45. M. DePaz, J. J. Leventhal, and L. Friedman, *J. Chem. Phys.* **51**, 3748 (1969).

46. C. Dekerckheer, O. Dahlem, and J. Reisse, *Ultrason. Sonochem.* **4**, 205 (1997).

47. G. S. Kell and E. Whalley, *J. Chem. Phys.* **62**, 3496 (1975).

48. J. C. Tanger, Jr. and K. S. Pitzer, *AIChE J.* **35**, 1631 (1989).

49. Y. Marcus, *Ion Solvation*, Wiley, Chichester, 1985.

50. I. M. Abdulagatov, V. I. Dvoryanchikov, and A. N. Kamelov, *J. Chem. Eng. Data*, **43**, 830 (1998).

51. A. I. Abdulagatov, G. V. Stepanov, I. M. Abdulagatov, A. E. Ramazanova, and G. S. Alisultanova, *Chem. Eng. Commun.* **190**, 1499 (2003).

2

BULK PROPERTIES OF SCW

In this chapter, a variety of measured physical properties of supercritical water (SCW) is presented and discussed. As SCW exists over a range of thermodynamic states, determined by any two quantities from among the temperature, the pressure, and the density, it is imperative to have a good grasp on the *PVT* states available for SCW. These are found in suitable tables and are expressed by equations of state (EoSs). From these EoSs other quantities, such as the heat capacity, enthalpy, entropy, fugacity, and so on, of SCW can be derived. Other important measurable physical quantities of SCW are the static relative permittivity, the light refraction, the electrical and thermal conductivities, the viscosity and diffusivity, and the ionic dissociation. These are dealt with in the following sections.

2.1 EQUATIONS OF STATE (EoS)

2.1.1 PVT Data for SCW

The density (mass per volume) of water, ρ, or the specific volume of water, v, that is, the volume for unit mass, has been carefully measured by many authors as a function of the temperature and pressure. The data are summarized in the Steam Tables [1–3] below the critical point, that is, for liquid water and water vapor, including values along the saturation line [4], where equilibria between liquid water and its vapor (vapor–liquid equilibria, VLE) prevail. The Steam Tables include also data above the critical point, that is, for supercritical water

Supercritical Water A Green Solvent: Properties and Uses, First Edition. Yizhak Marcus.
© 2012 John Wiley & Sons, Inc. Published 2012 by John Wiley & Sons, Inc.

(where no two-phase equilibria exist). Such data are called collectively pressure–volume–temperature, in short *PVT*, data.

Saul and Wagner [5] critically evaluated the *PVT* data over the temperature range from the melting line up to 1273K and for pressures up to 25 GPa and summarized them by means of a 58-parameter equation of state. Subsequently Wagner and Pruss [6] reported the IAPWS-95 formulation of the data (IAPWS: International Association for the Properties of Water and Steam). SCW did not receive any special attention in these extensive reports.

In order to have some feeling for the numbers involved, Table 2.1 shows a *very* abbreviated listing of water densities at temperatures and pressures characteristic of some processes in which SCW is being employed. The densities ρ_W are in $kg\,m^{-3}$, the temperatures t are in °C (T/K 273.15) and the pressures P are in MPa (1 MPa = 10 bar = 9.86923 atm). The tabulated densities at the (t, P) range in the table are seen to diminish considerably from their values at low temperatures (slightly above the critical temperature, $\approx$374°C) and high pressures, ca. 629 $kg\,m^{-3}$, to those at high temperatures and low pressures (somewhat above the critical pressure, $\approx$22.1 MPa), about 58 $kg\,m^{-3}$ and lower still at higher temperatures. However, densities outside the tabulated range, both larger and smaller are readily available for SCW. When $\rho_W < 200\,kg\,m^{-3}$ the SCW is said to exist in a gas-like state, whereas at larger densities it exists in a liquid-like state, although in both cases it is a supercritical fluid that does not distinguish between gas and liquid.

Accurate critical point data for the water substance [6] are shown in Table 1.2. Note that the critical density is known only to ±1%, hence so is

TABLE 2.1 The Densities ρ/kg m^{-3} of SCW at Some Thermodynamic States (Data from Ref. [45])

t/°C	P/MPa								
	25	30	35	40	50	60	70	80[a]	100[a]
380	446.40	533.70	570.00	594.60	629.37	654.5	674.8	693.0	721.3
400	166.28	353.28	473.75	523.81	578.34	612.7	636.1	660.3	693.4
450	109.04	148.70	201.78	272.12	401.27	479.1	528.0	563.0	614.0
500	89.86	115.19	144.40	178.06	257.60	338.8	405.3	459.8	528.2
550	78.61	98.37	119.88	143.23	195.58	266.3	303.7	371.0	449.5
600	70.79	87.44	105.05	123.64	163.64	207.2[a]	251.7[a]	295.5	370.8
650	64.87	79.48	94.68	110.46	143.68	179.1[a]	215.4[a]		
700	60.13	73.28	86.81	100.70	129.53	159.8[a]	190.7[a]		
750	56.21	68.24	80.52	93.04	118.76	145.3[a]	172.4[a]		
800	52.89	64.02	75.33	86.80	110.18	134.0[a]	158.3[a]		

[a] Data from Ref. [67].

the critical molar volume, $V_c = 56.0 \, \text{m}^3 \, \text{mol}^{-1}$, too. The critical compressibility factor of water is $Z_c = P_c V_c / RT_c = 0.229$ and differs (is lower than) the value of Z_c for conformal liquids, 0.291.

The isobaric expansibility of SCW, $\alpha_p = V^{-1}(\partial V/\partial T)_P$, and its isothermal compressibility $\kappa_T = V^{-1}(\partial V/\partial P)_T$, can be derived from the tabulated values but are also obtained from the equations of state (Section 2.1.2).

2.1.2 Classical Equations of State of SCW

The modeling of SCW by means of equations of state with fewer parameters than the 58 used by Saul and Wagner [5] is expedient. The EoS generally describes the compressibility factor:

$$Z = PV/RT \qquad (2.1)$$

and represents the independent variables by means of the reduced quantities: the reduced temperature $T_r = T/T_c$ and the reduced pressure $P_r = P/P_c$. The compressibility factor then yields the reduced molar volume $V_r = V/V_c$ or the reduced density $\rho_r = \rho/\rho_c$. Alternatively, the reduced volume or density can play the role of an independent variable, yielding the reduced pressure.

Duan et al. [7] reported a 15-parameter EoS for SCW that is valid up to 1000°C and 3500 bar (350 MPa) and describes the experimental data well (within 2%):

$$Z = \Sigma a_{ij} T_r^{-i} V_r'^{-j} + b_1 T_r^{-3} V_r'^{-2}(b_2 + b_3 V_r'^{-2})\exp(-b_3 V_r'^{-2}) \qquad (2.2)$$

where $V_r' = VP_c/RT_c$ (*not* V/V_c). Note that not all the possible i and j values are used; the values of the coefficients a_{ij} and b_i employed are shown in Table 2.2.

TABLE 2.2 Coefficients of the Duan et al. EoS [7]

Parameter	Value	Parameter	Value
a_{00}	1	a_{24}	0.00015521
a_{01}	0.0864449	a_{34}	−0.00010686
a_{21}	−0.3969190	a_{05}	−0.000004932
a_{31}	−0.0573335	a_{25}	−0.000002737
a_{02}	−0.0002939	a_{35}	0.000002656
a_{22}	−0.0041578	b_1	0.00896079
a_{32}	0.0199497	b_2	4.02
a_{04}	0.00011890	b_3	0.0257

Many cubic EoSs have been found to be useful for the description of *PVT* data of fluids, including supercritical ones. Kerrick and Jacobs [8] modified the repulsive term of the Redlich–Kwong EoS in order to model SCW, by leaning on scaled particle theory, but retained the attractive term. The expression used was

$$Z = (1 + y + y^2 - y^3)(1 - y^3)^{-1} - a(V, T)T^{-3/2}(V + b)^{-1} \qquad (2.3)$$

The covolume $b = 29\,\text{cm}^3\,\text{mol}^{-1}$ represents close-packed hard spheres, to which also the packing parameter $y = b/4V$ refers. The a parameter has the form $c + dV^{-1} + eV^{-2}$ and the coefficients c, d, and e are quadratic polynomials in T, requiring six parameters in all. This EoS described the *PVT* data of SCW well up to 925°C and 840 MPa.

Kutney et al. [9] proposed a so-called volume-translated version of the Redlich–Kwong EoS modified in terms of the scaled particle theory for hard sphere particles. First the pressure is expressed in terms of the untranslated volume V^{ut} as

$$P = (RT/V^{\text{ut}})(V^{\text{ut}\,3} + bV^{\text{ut}\,2} + b^2 V^{\text{ut}} - b^3)/(V^{\text{ut}} - b)^3 - a/(V^{\text{ut}} + 2b)^2 \qquad (2.4)$$

From this the implicit V^{ut} is extracted with $b = 0.02459RT_c/P_c$ and $a = 0.4495$ $\alpha R^2 T_c^2/P_c$, and $\alpha = \exp[(1 - T_r)(\alpha_A T_r^{-0.93} + \alpha_B T_r^{0.75})]$. Then the translated volume is obtained from

$$V = V^{\text{ut}} + t + (V_c - V_c^{\text{ut}} - t)(8\,V_r^{\text{ut}}\,T_r^{-4.5})/(V_r^{\text{ut}\,3} + 6.5\,T_r^{-6.5} + 0.5) \quad (2.5)$$

Here $V_r^{\text{ut}} = 0.3184RT_c/P_c$ and the various numerical coefficient are universal, that is they apply to all the pure substances studied. There remain three substance-specific parameters that for water are $t = 0.0048$, $\alpha_A = 0.105$, and $\alpha_B = 1.038$. The set of equations and parameters described the EoS of water well from ambient to 600°C and 30 MPa.

Chern [10] fine-tuned the empirical Peng–Robinson cubic EoS, but required a modified acentric factor namely $\omega = 0.5491$, rather than the normal value for water $\omega = 0.3443$. The resulting 3-parameter expression for the compressibility factor Z is

$$Z = (1 - b/V)^{-1} - 1.115 \times 10^{-4}T_r^{-1}\,[1 + 1.140(1 - T_r^{1/2})]^2/[V + 2b - b^2/V] \qquad (2.6)$$

where b is the covolume parameter, but its value was not specified in the paper. This expression was useful in the range 400–600°C and up to 60 MPa.

Such EoSs as Eqs. (2.2)–(2.6), though able to model the *PVT* data (and other kinds of data, see below) of SCW with a limited number of parameters, are not derived from any theory regarding the properties of supercritical water, which, in turn, are related to models describing the interactions prevailing in SCW, mainly the hydrogen bonding.

An EoS of SCW, based on measurements of ultrasound velocities, relevant to geochemical conditions prevailing in the earth's crust at 400°C but at very high pressures, up to 5.5 GPa, was reported by Abramson and Brown [11] and its consequences were confirmed by a theoretical calculation by Bastea and Fried [12].

2.1.3 Scaling Equations of State for SCW

Several theoretical approaches, based on models that take the hydrogen bonding still present in SCW explicitly into account, have been the basis of proposed EoSs. It this book attention is given to thermodynamic states not in the immediate vicinity of the critical point, but rather to values ranging from somewhat beyond it. If the region near the critical point is of interest, the papers by Anisimov et al. [116], Abdulagatov et al. [13], and Smits et al. [14] should be consulted.

A lattice fluid hydrogen bonding (LFHB) model was proposed by Gupta et al. [15]. It is based on the Panayiotou–Sanchez EoS [16] and is applicable to hydrogen bonded fluids such as alcohols and water. It yields the EoS as

$$\rho_r''^2 + P_r'' + T_r'' \left[\ln(1 - \rho_r'') + \rho_r'' \left(1 - 1/r\right)\right] = 0 \qquad (2.7)$$

Here the scaling of the reduced quantities is *not* with regard to the critical ones, but $\rho_r'' = \rho/\rho^*$ (presumably, but not so defined in the paper), $P_r'' = P/P^*$, and $T_r'' = T/T^*$. For water the scaling parameters are $\rho^* = 853 \, \text{kg m}^{-3}$, $P^* = 475 \, \text{MPa}$, and $T^* = 518 \text{K}$. The segment length, r, is a dimensionless quantity not clearly specified in the paper. The resulting chemical potential can be rewritten as

$$\mu/RT = r\{(-\rho_r'' + PV_r'')/T_r'' + V_r'' - 1) \ln\left(1 - \rho_r''\right) + (1/r) \ln \rho_r''\} \\ + (2 - 2f_0) + 4 \ln f_0$$

$$(2.8)$$

where $V_r'' = 1/\rho_r''$. It yields the fraction f_0 of monomeric water (not hydrogen bonded at all) as

$$f_0 = [(A^2 + 8A)^{1/2} - A]/4 \qquad (2.9)$$

where the parameter A depends on the Gibbs energy for hydrogen bond formation:

$$A = rV_r'' \exp(G°) = rV_r'' \exp[E° - TS° + PV°]/RT \qquad (2.10)$$

The hydrogen bonding parameters used are $E° = -15.5 \, \text{kJ mol}^{-1}$, $S° = -16.6 \, \text{J K}^{-1} \, \text{mol}^{-1}$, and $V° = -4.2 \, \text{cm}^3 \, \text{mol}^{-1}$. Six parameters are thus specified: ρ^*, P^*, T^*, $E°$, $S°$, and $V°$ and the unspecified r to describe quantitatively the PVT values and the chemical potential of supercritical water, as well as the derived fraction of monomeric water f_0. The latter is shown in a figure for three temperatures, 400, 600, and 800°C and $0 \leq P/$ MPa≤ 100.

In a subsequent paper Gupta and Johnston [17] improved the LFHB model, leaning on the statistical associated fluid theory (SAFT) of Chapman et al. [18] and Huang and Radosz [19]. Rather then the empirical ρ^*, P^*, T^* the improved model employs a segment volume $v°° = 10 \, \text{cm}^3 \, \text{mol}^{-1}$ and an energy parameter $u_0/k_B = 218.80 \text{K}$, with the parameter $r = 1.33$, obtained by fitting water data up to the critical point. The hydrogen bonding parameters used differ somewhat from those previously employed: $E° = -15.47 \, \text{kJ mol}^{-1}$, $S° = -15.65 \, \text{J K}^{-1} \, \text{mol}^{-1}$, and $V° = -0.97 \, \text{cm}^3 \, \text{mol}^{-1}$. The density of SCW at 450°C at pressures up to 80 MPa is modeled well by the new approach.

Smits et al. [14] preferred the associated perturbed anisotropic chain theory (APACT) model of Ikonomou and Donohue [20] with three sites of hydrogen bonding per water molecule [21] for the EoS at $200 \leq t/°C \leq 1000$ and $2.5 \leq P/\text{MPa} \leq 5000$. The theory requires four molecular parameters, obtained from fitting liquid water data up to the critical point, in order to model the data for SCW. Smits et al. [14] did not present the expressions for obtaining Z or μ (cf. [20, 21]) but did the fitting parameters: a characteristic energy, $T^* = 183.8 \text{K}$ and a characteristic volume $v^* = 11.64 \, \text{cm}^3 \, \text{mol}^{-1}$, as well as the enthalpy $\Delta H° = -20.10 \, \text{kJ mol}^{-1}$ and entropy $\Delta S° = -10.97 \, \text{J K}^{-1} \, \text{mol}^{-1}$ yielding the equilibrium constant for hydrogen bonding $\Delta = \exp(-\Delta H°/RT + \Delta S°/R)$. Smits et al. [14] showed good agreement between experimental and calculated PVT data. The fraction of monomeric (not hydrogen bonded) water molecules is

$$f_0 = 2/[1 + 4\rho\Delta - (\rho\Delta)^2 + (1 + \rho\Delta)\{1 + 6\,\rho\Delta + (\rho\Delta)^2\}^{1/2}] \qquad (2.11)$$

They compared values of f_0 from their approach with values from the earlier LFHB approach of Gupta et al. [15] (but not the improved one [17]) and the SAFT of Huang and Radosz [19], finding considerable differences. This subject is elaborated on in Chapter 3.

Touba and Mansoori [22] applied the analytic chain association theory (ACAT) to sub- and supercritical water. The fluid is composed of an equilibrium mixture containing monomers, dimers, and so on, and an EoS is derived permitting the calculation of the *PVT* values as well as the radial distribution function (Chapter 3). This was applied to SCW at 400°C and a density of $660 \, \text{kg m}^{-3}$ in good agreement with experimental values.

Abdulagatov et al. [13] applied the simplified perturbed hard chain theory (SPHCT) for water, including SCW. The EoS from this theory is

$$Z = 1 + c(4\tau V_r'^{-1} - 2\tau^2 V_r'^{-2})(1 - \tau V_r'^{-1})^{-3} - Z_m cY/(V_r' + Y) \quad (2.12)$$

Here $\tau = 0.7405$ is the geometrical factor for close packing of spheres, $Z_m = 18$ is the maximal coordination number, $Y = \exp(1/2T'_r) - 1$, and the reduced quantities are $T'_r = T/T^*$ and $V'_r = V/V^*$. There are three fitting parameters $c = 1.021$, $T^* = 1315 \text{K}$, and $V^* = 12.8 \, \text{cm}^3 \, \text{mol}^{-1}$; the latter is the molar close-packed volume, close to the van der Waals volume of water, $12.4 \, \text{cm}^3 \, \text{mol}^{-1}$. For SCW the fitting was only up to 420°C at rather low pressures (density near the critical one). The main purpose of the calculations was the obtaining of the isochoric heat capacity (see Section 2.2).

In a later development of the LHFB approach Vlachou et al. [23] fixed the segment number for water at $r = 1$ and used different values for the hydrogen bonding parameters: $E° = -19.9 \, \text{kJ mol}^{-1}$, $S° = (-)26.5 \, \text{J K}^{-1} \, \text{mol}^{-1}$ (the minus sign was omitted in the paper) with a temperature and pressure dependent volume of hydrogen bonding:

$$V° = [-0.51 - 0.00203(P/\text{MPa})]$$
$$\cdot [1 - 5.66 \times 10^{-5}(T/K - 383.15)^3/|T/K - 373.15|] \quad (2.13)$$

The values of the scaling parameters are then $\rho^* = 1000 \, \text{kg m}^{-3}$, $P^* = 186.4$ MPa, and $T^* = 404 \text{K}$. The *PVT* data were well fitted between 380 and 610°C and 40–100 MPa. The resulting degree of hydrogen bonding within these ranges were shown in a figure, the fraction of monomeric water f_0 being calculated as specified above, Eq. (2.9). The treatment is simplified by limiting the model to the SCW conditions, though it takes into account two types of nonrandom distributions: that of the free volume and the orientation of the hydrogen bonding molecules.

Another recent development is the modified SAFT, that is, the crossover SAFT EoS of Hu et al. [24]. To the 5 SAFT parameters [19] were added 5 additional ones, and of the total number 7 were adjustable parameters, obtained from data fitting, and were listed. For water the calculated critical constants are much nearer the experimental values that those obtained from the classical SAFT calculations. *PVT* values were calculated by means of the crossover SAFT expressions but only within a narrow temperature range for SCW (up to 440°C) and at a nonspecified pressure range, and still with an average absolute deviation (AAD) of 4.3%.

Bermejo et al. [25] applied the Anderko–Pitzer EoS [26], originally contrived for $H_2O + NaCl$ mixtures at $300 \leq t/°C \leq 930$ and $P \leq 500 \, MPa$, to the SWC oxidation process, that is, to the behavior of the components of air (N_2 and O_2) in SCW. The EoS consists of a reference and a perturbation part, the former arising from the sum of a repulsion and a dipolar term and the latter is in the form of a virial expansion. In addition to the required acentric factor, $\omega = 0.3443$, the 14 parameters needed for water were fitted by Kosinski and Anderko [27], permitting the expression of the density at $100 \leq t/°C \leq 930$.

In summary, there exist a plethora of models that describe the EoS, that is, the *PVT* properties, of SCW, only some of which are shown here. They may have a better or not so good success, and this over a wider or more restricted range of thermodynamic states (temperature and pressure). Since the *PVT* data themselves are available (Section 2.1), the main purpose of the EoSs is to derive other quantities, such as the chemical potentials, isochoric heat capacities, the fractions of monomeric (not hydrogen bonded) water molecules, and so on. These are described more fully in the appropriate sections of this book.

2.1.4 EoS of Supercritical Heavy Water

The critical parameters for D_2O are $T_c = 643.89 \, K$ (370.74°C, 3.23K below that of H_2O), $P_c = 21.67 \, MPa$, and $\rho_c = 355.8 \, kg \, m^{-3}$.

Kiselev et al. [28] reported a parametric crossover model for the representation of the EoS of heavy water, D_2O. They fitted with an AAD (absolute average deviation) of 0.1% the *PVT* data of D_2O (with a small correction for an 0.2 mol% H_2O impurity [29]) in the ranges $0.98 \leq T_r \leq 1.10$ and $0.5 \leq \rho_r \leq 1.5$, resulting in 14 parameters. Mixtures of $H_2O + D_2O$ were also treated in this paper. The critical data of the mixtures depend linearly on the mole fraction composition, x_{H_2O} [30]. The EoS was able to represent the thermodynamic data for the mixtures in the ranges $0.8T_c(x) \leq T \leq 1.5T_c(x)$ and $0.35\rho_c(x) \leq \rho \leq 1.65\rho_c(x)$. In a later paper, Bazaev et al. [31] presented experimental *PVTx* data at $x_{H_2O} = 0.5$ and 0.6 in the near-critical and supercritical regions up to 680K and 37.5 MPa, that is, not far from the critical region, in good agreement with the model of Kiselev et al. [28].

2.2 THERMOPHYSICAL PROPERTIES OF SCW

2.2.1 Heat Capacity

The ideal gas molar heat capacity of water at constant pressure, C_p^{ig} was calculated by Whoolley from spectroscopic data by means of statistical thermodynamics [32]. His earlier reported values were modeled by a 10-parameter equation by Cooper [33]. For the temperature range of interest for supercritical water, $C_p^{ig}/J \cdot K^{-1} \cdot mol^{-1}$ increases steadily from 36.90 at 650K to 44.36 at 1250K, that is, with a slope of 0.0125 $J K^{-2} mol^{-1}$. The isochoric molar ideal gas heat capacity of water was reported by Vargaftig [34] as a quadratic in the temperature:

$$C_v^{ig}/J K^{-1} mol^{-1} = 25.85 - 0.3325 \times 10^{-2}(T/K) + 0.6906 \times 10^{-5}(T/K)^2$$

$$(2.14)$$

resulting in 26.60 at 650K and 32.49 at 1250K.

The actual specific heat at constant pressure of the water substance, c_p, at supercritical conditions, up to 700°C and 100 MPa was reported by Sirota and Belyakova [35]. The values tabulated in the Steam Tables [2] up to 2000°C and 3 GPa are based on these data, and a brief survey of the data is shown in Table 2.3. At given pressures the molar heat capacity C_p (the molar mass of water times the specific heat) has a maximum at increasing temperatures and then decreases slowly up to very high values of these (> 1000°C). There new

TABLE 2.3   The Molar Heat Capacity at Constant Pressure, $C_p/J K^{-1} mol^{-1}$, of SCW at Some Thermodynamic States (Data from Ref. [2])

$t/°C$	*P*/MPa						
	25	30	35	40	50	60	70
380	421.1	177.4	141.5	125.3	109.2	100.7	95.28
400	238.3	451.8	210.3	157.0	123.9	108.3	100.4
450	92.04	123.2	168.4	201.9	169.4	135.4	117.1
500	67.32	77.68	90.18	104.5	130.4	135.7	126.0
550	58.11	63.74	70.02	76.86	91.07	102.5	107.7
600	53.54	56.93	60.76	64.80	73.18	81.07	87.30
650	50.71	53.25	55.92	58.66	64.26	69.07	74.49
700	49.16	51.09	53.05	55.07	59.14	63.11	66.75
750	48.24	49.76	51.29	52.84	55.95	59.00	61.85
800	47.72	48.93	50.17	51.49	53.90	56.31	58.60

internal modes of excitation of individual water molecules (vibrational levels) become available and C_p rises slowly again. At given temperatures the pressure dependence C_p again shows maxima at moderate to high pressures that become shallower at increasing temperatures.

If the heat capacity density, $[C_p - C_p^{ig}]/V$, is used as a criterion for structuredness of a fluid [36] the data in Tables 2.1 and 2.3 can be combined to yield these values, shown in Fig. 2.1 for three pressures as functions of the temperature. The criterion for structuredness is $[C_p - C_p^{ig}]/V \geq 0.6$ J K^{-1} cm^{-3} and SCW is seen to be structured at low temperatures, the threshold being crossed at higher temperatures the higher the pressure. This is understandable in terms of the interactions between the water molecules, as described in Chapter 3.

The Sirota and Maltsev [37] heat capacity data near the critical point have been challenged by Levelt Sengers et al. [38]. They found a small offset of the peak values of C_p for isobars between 22.56 and 26.97 MPa when plotted against the densities and suggested a lowering of the temperatures reported by Sirota and Maltsev by 0.05K to correct this. This suggestion was accepted by Wagner and Pruss [6].

The isochoric heat capacity, C_v, of SCW and of near-critical water have been studied intensively by the group of Abdulagatov [13, 39–42]. Abdulagatov et al. [13] applied the SPHCT for water, including SCW, to model their

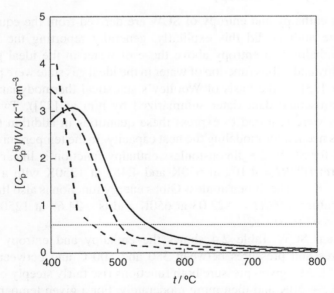

FIGURE 2.1 The heat capacity density, $[C_p - C_p^{ig}]/V$, of SCW as a function of the temperature at three pressures: (- - -) 30 MPa, (– – –) 40 MPa, and (—) 50 MPa. The dotted line represents the threshold of "structuredness" according to this criterion.

data. On the basis of the EoS from this theory, Eq. (2.12), the isochoric heat capacity is

$$C_v = C_c^{ig} + RZ_m c(V_r' - 1)(Y + 1)(V_r' + Y)^{-2}(2T_r')^{-2} \qquad (2.15)$$

The meaning of the quantities is given above, following Eq. (2.12).

The isobaric heat capacity of supercritical heavy water was measured by Rivkin and Egorov [43] up to 30 MPa and 450°C. The isochoric heat capacity of supercritical heavy water was studied by Abdulagatov et al. [13] up to 110 MPa and 473°C, including the vicinity of the critical point: $0.97 \leq T_r \leq 1.03$. Polikhrocidi et al. [44] studied the isochoric heat capacity of equimolar $H_2O + D_2O$ in the critical region.

Closely related to the heat capacities of SCW is the speed of sound in it (Section 2.2.3), because it yields, together with the densities, the isentropic compressibility: $\kappa_S = 1/\rho u^2$. The ratio of the isobaric compressibility to the isentropic one equals the ratio of the isobaric and isochoric heat capacities: $\kappa_T/\kappa_S = C_p/C_v$ (note that $\kappa_T = \rho^{-1}(\partial\rho/\partial P)_T$). Therefore, one of these four quantities, when not measured, can be calculated from the other three.

2.2.2 Enthalpy and Entropy

The molar enthalpy and entropy of SCW are derived from the equations of state. Some authors did this explicitly, generally reporting the so-called residual enthalpy and entropy above those of water in the ideal gas state. The enthalpy and Gibbs function of water in the ideal gas state were presented by Cooper [33] on the basis of Woolley's statistical thermodynamic study from spectroscopic data (later summarized by him as [32]). Two further parameters were required to express these quantities in addition to the 11 parameters needed for modeling the heat capacity. In the temperature interval of interest for SCW, the dimensionless enthalpy function is linear with the temperature: $H^{ig}/RT = 4.102$ at 650K and 4.483 at 1250K with a slope of $6.35 \times 10^{-4}\,K^{-1}$. The dimensionless Gibbs energy function is also linear with the temperature: $G^{ig}/RT = -22.03$ at 650K and is -24.68 at 1250K with a slope of $-4.41 \times 10^{-3}\,K^{-1}$.

For actual SCW Table 2.4 shows the enthalpy and entropy at some temperatures and pressures between 380 and 800°C and between 25 and 50 MPa [45]. At a given pressure both functions rise fairly steeply beyond T_c up to ca. $T_c + 50$K and then more moderately. For a given temperature the values of both functions diminish somewhat with increasing pressures. Detailed values of these functions in the near critical region along isochors

TABLE 2.4   The Molar Enthalpy H/kJ mol^{-1} and Entropy S/J K^{-1} mol^{-1} of SCW at Some Thermodynamic States (Data from Ref. [45])

$t/°C$	P/MPa							
	25		30		40		50	
	H	S	H	S	H	S	H	S
380	34.97	75.23	33.11	72.10	32.00	69.92	31.47	68.66
400	46.51	92.70	38.94	80.88	34.85	74.22	33.83	72.21
450	53.22	102.37	50.90	98.17	45.32	89.19	41.31	82.92
500	57.03	107.47	55.58	104.44	52.37	98.65	49.05	93.29
550	60.12	111.33	59.04	108.79	56.78	104.19	54.43	99.67
600	62.87	114.58	62.03	112.31	60.29	108.33	58.52	104.86
650	65.46	117.46	64.76	115.36	63.36	111.76	61.95	108.69
700	68.09	120.10	67.37	118.11	66.20	114.76	65.04	111.94
750	70.40	122.55	69.90	120.65	68.92	117.48	67.93	114.84
800	72.81	124.85	72.39	123.02	71.55	119.99	70.71	117.50

up to 695 K ($\sim T_c + 50$) and 420 kg m^{-3} (40.5 MPa) were presented by Levelt Sengers et al. [38].

The equations of state of SCW (Sections 2.1.2 and 2.1.3) permit the calculation of the molar Helmholtz and Gibbs energies, A and G, hence the chemical potentials μ, fugacities f, molar enthalpies H, and entropies S. Duan et al. [7] calculated the fugacity from the compressibility factor Z (Eq. (2.2))):

$$\ln(f/P) = Z - 1 - \ln Z - \int_{\infty}^{V_r'} [(P_r/T_r) - (1/V_r')]dV_r' \qquad (2.16)$$

They presented a table with the fugacity coefficients f/P of SCW for $400 \leq t/°C \leq 1200$ and $300 \leq P/\text{MPa} \leq 1000$. Kutney et al. [9] used their volume-translated modified Redlich–Kwong EoS (Eqs. (2.4) and (2.5)) to calculate the residual enthalpies and entropies of SCW as

$$H^r = H - H^{\text{ig}} = PV - RT - \int_{\infty}^{V} [P - T(\partial P/\partial T)_V]dV \qquad (2.17)$$

$$S^r = S - S^{\text{ig}} = R \ln Z - \int_{\infty}^{V} [(R/V) - (\partial P/\partial T)_V]dV \qquad (2.18)$$

They presented figures comparing the calculated residual enthalpies with those derived from the Steam Tables [45] with fair agreement. Anderko and

Pitzer [26] presented the residual Helmholtz energy as a sum of a reference term and a perturbation term, the former being the sum of a repulsion term and a dipolar interaction term. Bermejo et al. [25] reproduced the specific enthalpy of SCW at 30 MPa, which agreed well with the Anderko–Pitzer EoS [26] up to 480°C. Kretzschmar et al. [46] presented supplementary backward expressions for the calculation of two of the variables pressure, density, and temperature of SCW given one of them and the enthalpy or the entropy.

2.2.3 Sound Velocity

The velocity of ultrasound, u/m s^{-1}, in SCW was reported by Petitet et al. [47] up to 700°C and from 50 to 300 MPa. The velocity diminishes as a second power expression with the temperature at constant pressure and increases similarly with the pressure at a constant temperature. A summarizing equation can be expressed as follows:

$$
\begin{aligned}
u/\text{ms}^{-1} = &\, [4019 - 8.21(P/\text{MPa}) + 0.01036(P/\text{MPa})^2] \\
&- [13.17 - 0.0280(P/\text{MPa}) + 7.6 \times 10^{-5}(P/\text{MPa})^2](t/°\text{C}) \\
&+ [0.01161 - 5.966 \times 10^{-5}(P/\text{MPa}) \\
&+ 8.475 \times 10^{-8}(P/\text{MPa})^2](t/°\text{C})^2
\end{aligned}
$$

$$(2.19)$$

At 380°C u/m s^{-1} rises from 848 at 50 MPa to 1695 at 300 MPa and at 100 MPa it diminishes from 1153 at 380°C to 794 at 700°C, but it exhibits a small minimum within the range of data at 100 MPa and 628°C. The ultrasound data at 400°C were more recently extended by Abramson and Brown [11] up to very high pressures, 5.5 GPa.

The reciprocal of the product of the square of the sound velocity and the density yields the adiabatic compressibility: $\kappa_S = 1/\rho u^2$. The resulting κ_S /GPa^{-1} values at 380°C are 2.98 at 50 MPa and 1.68 at 100 MPa, but at 100 MPa they increase with the temperature to 1.22 at 400°C and to 4.64 at 500°C.

2.3 ELECTRICAL AND OPTICAL PROPERTIES

2.3.1 Static Relative Permittivity

There have been only few determinations or estimates of the static relative permittivity, ε_r, of water that extended into the region of SCW [48–50] prior to the definitive work of Heger et al. [51]. These authors presented experimental data up to 550°C and 500 MPa. Franck [48] and Quist and Marshall [49] had also previously presented a method based on the Kirkwood dipole orientation parameter g (Section 1.4, Eqs. (1.11) and (1.12)) for predicting the relative permittivity up to 800°C that yielded values in agreement with the experimental ones in the range of their availability to $\pm 4\%$. Uematsu and Franck [52] subsequently presented a 10-parameter expression in the reduced temperature $(T_r' = T/298.15\text{K})$ and density $(\rho_r' = \rho/1000 \text{ kg m}^{-3})$ to model the data in the above specified range, not restricted to SCW. A different 10-parameter expression was proposed by Archer and Wang [53], again for a wide temperature range, including SCW. The temperature range for the calculated relative permittivity was extended to 650°C and 1 GPa by Mulev and Smirnov [54] and to 1000°C and the pressures to 1.6 GPa by Franck et al. [55].

The calculations are based on the use of the Kirkwood dipole orientation parameter g, Eq. (1.12) being reformulated as the Kirkwood–Fröhlich equation:

$$(\varepsilon_r - \varepsilon_\infty)/(2\varepsilon_r + \varepsilon_\infty)/\varepsilon_r(\varepsilon_\infty + 2) = (4\pi N_A \rho g \mu^2/3\varepsilon_0 k_B T M_W) \qquad (2.20)$$

In the earlier estimates [48, 49] the infinite frequency permittivity was set as $\varepsilon_\infty = 1$. The product $g\mu^2$ was fitted to data for liquid water under pressure and the values were extrapolated to the SCW region with the same parameters. The dipole orientation parameter g for water was fitted to the temperature and density by Pitzer [56] by the expression

$$g = 1 + 2.68\rho + 6.69\rho^5[(565/T)^{0.3} - 1] \qquad (2.21)$$

The resulting values for the relative permittivity were considered an improvement over those yielded by the equation of Uematsu and Franck [52].

The most definitive recent report on this subject is that of Fernandez et al. [57], though the temperature was limited to 600°C but the pressure included values to 1.2 GPa. They based their calculation on the following modification of the above expression:

$$(\varepsilon_r - 1)/(\varepsilon_r + 2) = (N_A\rho/3)[\alpha/\varepsilon_0 + g\mu^2/3\varepsilon_0 k_B T\{9\varepsilon_r/(2\varepsilon_r + 1)(\varepsilon_r + 2)\}] \qquad (2.22)$$

In this expression $\alpha = 1.636 \times 10^{-40}\,\text{C}^2\,\text{J}^{-1}\,\text{m}^{-2}$ is the mean molecular polarizibility and $\mu = 6.138 \times 10^{-30}\,\text{C m}$ is the dipole moment of an isolated water molecule. Eq. (2.21) was rearranged into a quadratic equation in ε_r implicit in g that in turn is obtained from ε_r as

$$g = (2 + 1/\varepsilon_r)(k_B T/3\,\mu^2)[(3\varepsilon_0/N_A\rho)\,(\varepsilon_r - 1) - \alpha(\varepsilon_r + 2)] \qquad (2.23)$$

The quantity g was then modeled by means of a 14-parameter expression in the reduced temperature and density, which permitted the estimation of derivative functions with respect to the temperature and the pressure as well as the limiting slopes of the Debye–Hückel expression for the osmotic and activity coefficients of electrolyte solutions. The ε_r values for SCW interpolated from those of Fernandez et al. [57] are shown in Table 2.5.

The values diminish with increasing temperatures and increase with raised pressures. At the lower temperature they are of the order of moderately polar solvents but diminish toward those of nonpolar solvents at $\geq 600°C$, beyond which they are commensurate with those arising from the atomic and electron polarization by the electric field (expressed as the square of the refractive index, n^2, see Section 2.3.3).

An important quantity in these expressions is the electronic polarizability, $\alpha = 18.1458 \times 10^{-30}\, m^3$ that should account for the nonorientational contribution to ε_r. However, other values of α were suggested, Uematsu and Franck [52] used $1.44 \times 10^{-30}\, m^3$ and still other values have temperature, pressure, and frequency dependence of n^2 taken into account [58].

The Debye–Hückel limiting law coefficients were derived by Fernandez et al. [57], being dependent on the density and relative permittivity at a given temperature and pressure. The equation for the coefficient of the activity coefficient ($\ln \gamma_\pm = -A_\gamma I^{1/2}$) is

$$A_\gamma = (2\pi N_A \rho M_W)^{1/2} (e^2 / 4\varepsilon_0 \varepsilon_r k_B T)^{3/2} \qquad (2.23a)$$

TABLE 2.5 The Relative Permittivity, ϵ_r, of SCW at Some Thermodynamic States (Data from Ref. [57])

$t/°C$	P/MPa				
	30	40	50	60	70
380	10.49	13.09	14.27	15.10	16.01
400	6.05	9.62	11.77	13.10	13.80
450	2.18	4.41	6.49	8.25	9.37
500	1.78	2.41	3.46	4.93	6.28
550	1.60	1.96	2.52	3.19	4.24
600	1.50	1.79	2.12	2.46	3.00
650	*1.42*	*1.63*	*1.90*	*2.15*	*2.34*
700	*1.36*	*1.50*	*1.74*	*1.98*	*2.05*
750	*1.32*	*1.44*	*1.64*	*1.85*	*1.97*

Values outside of the range of experimental data are in *italics*.

Values were shown for SCW at 100 and 1000 MPa, the expression for 100 MPa is

$$A_\gamma = 11.931 - 0.0566(t/^\circ C) + 8.67 \times 10^{-5}(t/^\circ C)^2 \qquad (2.23b)$$

but those at 1000 MPa are only 0.5–0.3 times as large (because the temperature is increased).

Wasserman et al. [59, 60] modeled the static relative permittivity of SCW by means of molecular dynamics (MD) computer simulations, based on the SPC/E model of a water molecule. They found good agreement with experimental data for that part of the range studied, temperatures up to 1005°C, pressures up to 2000 MPa, and densities in the range $257 \leq \rho/\text{kg m}^{-3} \leq 1.111$, where they were available.

The permittivity of SCW as a function of the frequency of the electrical field was first studied by Okada et al. [61] by microwave spectroscopy. The appropriate expression from the Debye theory is

$$\varepsilon(\omega) = \varepsilon(\infty) + (\varepsilon(0) - \varepsilon(\infty))/(1 + i\omega\tau_D) \qquad (2.24)$$

Here ω is the frequency and τ_D is the dielectric relaxation time. The values of the static relative permittivity (at low frequencies, $\varepsilon(0)$) were taken from Uematsu and Franck [52] but assignment of the infinite frequency values, $\varepsilon(\infty)$, were problematic, so $\varepsilon(\infty) = 1$ at $\geq 200^\circ C$ was assumed, as for the dilute vapor. The resulting relaxation times had a minimum in the near-critical region, $\tau_D \sim 0.45$ ps, but increased toward 1 ps at $P = 59.4$ MPa and toward 2.7 ps at 35.5 MPa at the highest temperature examined, 750°C.

Further evaluation of the orientational relaxation times obtained from dielectric spectroscopic data is deferred to Section 3.3.2.

2.3.2 Electrical Conductivity

The specific electrical conductance of pure water at ambient conditions is, of course, very small: 5.5 µS m^{-1} at 25°C. However, it increases considerably as the temperature and pressure are increased and reaches appreciable values in SCW at not too low densities. The relevant data were obtained by Marshall [62] at up to 1000°C and 1000 MPa. The procedure followed was to obtain the limiting equivalent conductivities $\lambda^\infty(\text{H}^+)$ and $\lambda^\infty(\text{OH}^-)$ from experiments and the published [63] ion product of water $K_W = [\text{H}^+][\text{OH}^-]$ where the square brackets denote molalities (see Section 2.6) for the calculation of the specific conductivity of SCW. The limiting equivalent conductance isotherms of 1:1 electrolytes in SCW between 400 and 800°C was found to be a linear

TABLE 2.6 The Specific Conductance of SCW, $\kappa_W/\mu S\,m^{-1}$, According to Marshall [62]

	P/MPa				
$t/°C$	50	100	200	400	600
400	117	491	1430	3920	7130
600		13.4	465	3380	8570
800		0.694	83.1	1780	6540
1000		0.140	19.8	820	4160

function of the density. The combination of the data for NaCl, HCl, and NaOH then yields

$$\Lambda^\infty(H^+OH^-) = \Lambda^\infty(HCl) + \Lambda^\infty(NaOH) - \Lambda^\infty(NaCl) \qquad (2.25)$$

The resulting value valid for densities $400 \leq \rho/kg\,m^{-3} \leq 1000$ is

$$\Lambda^\infty(H^+OH^-)/S\,m^2\,mol^{-1} = 0.1850 - 0.131(\rho/kg\,m^{-3}) \qquad (2.26)$$

The specific conductance of SCW is then

$$\kappa_W = \Lambda^\infty(H^+OH^-)\,K_W^{1/2}\rho \qquad (2.27)$$

Representative values of κ_W resulting from this treatment are shown in Table 2.6.

2.3.3 Light Refraction

The light refraction by SCW has been reported by Schiebener et al. [64] up to 500°C and 100 MPa over a wide wavelength range. The square of the value at 200 nm, the lowest wavelength (highest frequency) studied, n_{200}^2, can be used as an approximation for the value at infinite frequency, n_∞^2, which in turn can be used as the value of the atomic + electronic permittivity, that is, the relative permittivity at infinite frequency $\varepsilon(\infty)$ in Eqs. (2.20) and (2.24). Table 2.7 shows the data interpolated from those of Schiebener et al. [64].

Kuge et al. [117] more recently reported the refractive index of SCW using a helium–neon laser up to 420°C and 27 MPa. The results show the refractive index n to be linear with the density under these conditions:

$$n = 1 + 3.3 \times 10^{-4}\,(\rho/kg\,m^{-3}) \qquad (2.28)$$

TABLE 2.7   The Square of the Refractive Index of SCW at 200 nm, n_{200}^2 (Data from Ref. [64])

$t/°C$	P/MPa					
	25	30	40	50	60	70
400	1.141	1.306	1.444	1.500	1.538	1.558
450	1.075	1.098	1.177	1.261	1.315	1.366
500		1.069	1.124	1.178	1.238	1.266

Inversion of the Lorenz–Lorentz relation yields the refractive index at infinite frequency from the polarizability of a water molecule α (Table 1.4) as

$$n_\infty = [(V + 8\pi N_A\alpha/3)/(V - 4\pi N_A\alpha/3)]^{1/2} \qquad (2.29)$$

where V is the molar volume of the SCW.

2.4   TRANSPORT PROPERTIES

2.4.1   Viscosity

The dynamic viscosity, η, of SCW was measured by several authors but the definitive study of Dudziak and Franck [65] for 400–560°C and 50–350 MPa summarized the previous results as well as their own. A sixth power polynomial in the density was required, of which only the coefficient of the ρ^6 terms was temperature dependent: $\eta = \sum a_i\rho^i$ with $a_6 = b_1 + b_2/T$. Up to a density of 800 kg m^{-3} the viscosity increased mildly with increasing temperatures (characteristic of the behavior of vapors) but beyond this density the viscosity diminished with increasing temperature, as expressed by the $a_6\rho^6$ term (characteristic of the behavior of liquids). Watson et al. [66] gave their attention mainly to the near-critical region (Section 1.5), disregarded the data of Dudziak and Franck [65], but presented tables of η of SCW at $375 \leq t/°C \leq 800$ and pressures up to 100 MPa. The Steam Tables [2] also presented detailed dynamic viscosity data for SCW at 400–800°C and 25–100 MPa, a representative list of which is shown in Table 2.8.

The IAPWS formulation for the viscosity of SCW is based on the critical examination of previous publications by Sengers and Kamgar-Paral [67], culminating in a 29-parameter equation covering the entire temperature and pressure ranges up to 900°C and 300 MPa. It, therefore, also includes the near-critical region where viscosity enhancement takes place. An alternative 19-parameter expression was also suggested for the entire range of

TABLE 2.8   The Dynamic Viscosity, η/μPa s, of SCW at Some Thermodynamic States (Data from Ref. [2])

$t/°C$	P/MPa						
	25	30	35	40	50	60	70
380	52.5[a]	61.3[a]					
400	29.0	43.8	55.8	61.3	67.9	72.4	76.0
450	28.9	30.8	34.0	39.1	50.7	58.2	63.8
500	30.6	31.7	33.2	35.2	40.7	47.4	53.4
550	32.5	33.4	34.4	35.7	38.9	43.0	47.4
600	34.5	35.2	36.0	37.0	39.3	42.1	45.3
650	36.5	37.1	37.7	38.6	40.4	42.6	45.0
700	38.4	38.9	39.6	40.2	41.8	43.6	45.5
750	40.3	40.8	41.3	41.9	43.3	44.8	46.4
800	42.1	42.6	43.1	43.7	44.9	46.2	47.6

[a]Data from Ref. [66].

thermodynamic states. Data published subsequent to this formulation were considered by Assael et al. [68], but no expression dealing specifically with SCW was given in either sources.

When attention is drawn to SCW, then a 5-parameter equation in the temperature and density expresses the dynamic viscosity sufficiently well:

$$\eta/\mu\text{Pa s} = [4.4668 + 0.04495(t/°C)]$$
$$+ [0.0131 + 2.89 \times 10^{-5}(t/°C)](\rho/\text{kg m}^{-3}) \qquad (2.30)$$
$$+ 9.2 \times 10^{-5}(\rho/\text{kg m}^{-3})^2$$

This equation is valid for $400 \leq t/°C \leq 800$ and pressures of $25 \leq P/\text{MPa} \leq 100$ (densities $50 \leq \rho/\text{kg m}^{-3} \leq 800$). In the range up to 600°C and similar pressures another expression with few parameters holds [58]:

$$\ln(\eta/\mu\text{Pa s}) = 1.04(1 - V_X/V) - 16,450/RT + 3.46g \qquad (2.31)$$

The first term takes care of the void space in SCW, $V_X = 16.7\ \text{cm}^3\ \text{mol}^{-1}$ being the intrinsic molar volume of water. The second term deals with the activation energy, and the dipole orientation correlation parameter g (Section 2.3.1) takes care of the hydrogen bonding that must be partly broken for the water to flow. Note that SCW has an apparent "negative activation energy" for viscous flow, contrary to liquid water. At low, gas like, densities of SCW its viscosity increases with increasing temperatures as expected from the Enskog theory [69]. For isochors at high densities, where SCW exhibits liquid-like properties, the viscosity still increases with increasing temperatures (contrary to expectations) but only mildly, because of the increasing pressure required to keep the density constant.

Rivkin et al. [70] measured the dynamic viscosity of heavy water at five temperatures from 400 to 500°C and at pressures up to 50 MPa. They found proportionality between the viscosity of heavy to light water throughout this range: $\eta(H_2O)/\eta(D_2O) = 1.32$.

2.4.2 Self-Diffusion

Although the viscosity of supercritical fluids is of the same order of magnitude as of gases at ambient conditions, the diffusivities are two orders of magnitude smaller (Table 1.2). The self-diffusion coefficient, D, of SCW was measured by Lamb et al. [71] using the NMR method for $400 \leq t/°C \leq 700$ and $0.315 \leq \rho_r \leq 2.21$ ($\rho_r = \rho/\rho_c$) obtained with $22 \leq P/MPa \leq 145$. At a density of 400 kg m^{-3} the values of $D/10^{-9} \text{ m}^2 \text{ s}^{-1}$ were 89.7 at 400°C, 95.6 at 500°C, 97.0 at 600°C, and 108 at 700°C and increased with diminishing densities. They could fit their results, within the experimental error of ±10%, to the empirical expression

$$\rho D = 2.24 \times 10^{-5} T^{0.763} \tag{2.32}$$

The largest deviations are at the low density end. The self-diffusion coefficients do not obey the Stokes–Einstein relationship for liquids

$$D = 2k_B T / C\pi\sigma\eta \tag{2.33}$$

with either a stick (6) or a slip (4) boundary condition C and a mildly varying diameter of the water molecules, $\sigma/\text{nm} = 0.275 - 2.2 \times 10^{-5}[(T/K) - 298.15]$ [72]. At states above the critical density the resulting $C = 2k_B T/D\pi\sigma\eta$ has temperature-dependent intermediate values between 4 and 6, whereas below this density the more gas-like Enskog theory behavior is followed.

Two dozen years later there appeared another report, by Yoshida et al. [73], of experimental data for the self-diffusion coefficient of SCW obtained by a high-temperature NMR probe, albeit at only one temperature, 400°C and rather low densities (~70 to ~400 kg m^{-3}). The results were in essential agreement with those of Lamb et al. [71] but with a much better claimed accuracy of ±1%. Furthermore, parallel measurements were carried out with heavy water, the ratio $D(H_2O)/D(D_2O) = 1.076$ at this temperature and practically independent of the density. Subsequently Yoshida et al. [74]

extended the measurements at 400°C to higher densities, 590–980 kg m^{-3}, D declining sharply with increasing densities.

A quasielastic neutron scattering method was employed by Beta et al. [75] for SCW at a rather narrow temperature range, $380 \leq t/°C \leq 450$ and pressures yielding fairly high densities of $578 \leq \rho/\text{kg m}^{-3} \leq 793$. An uncertainty of $\pm 10\%$ was associated with the results that were somewhat lower than those of Lamb et al. [71] at the common temperature, 400°C, of the two studies. Tassaing et al. [76] reported quasielastic neutron scattering results at 380°C in better agreement with the NMR data.

Tong et al. [77] employed the SAFT and a hard sphere model to study the self-association of water, including SCW with the data of Lamb et al. The theory requires three parameters: the hard sphere diameter of water (0.265 nm), the hydrogen bonding association energy ($e/k_B = 3675\text{K}$) and the association volume (5.8×10^{-6}, units not specified) and modeled the self-diffusion coefficient of not only SCW but also water along the saturation curve within 9% average absolute deviation.

Molecular dynamics computer simulations (Section 3.2.2) have been used by several authors for the determination of the self-diffusion coefficients. The direct result from the simulations is the velocity autocorrelation function of the oxygen atoms, which is translated into the self-diffusion coefficients. Kalinichev [78] calculated the values for five states at *nominal* temperatures between 357°C ($T_r = 1.03$) and 499°C ($T_r = 1.28$) and densities from 167 to 1284 kg m^{-3} in fair agreement with the experimental values of Lamb et al. [71], the Stokes–Einstein values of C in Eq. (2.33) ranged from 3.5 to 11.6. Mizan et al. [79] reported calculated self-diffusion coefficients for SCW at the *nominal* temperature of 500°C and at four densities: 115–659 kg m^{-3} in good agreement with the experimental values [71]. Yoshii et al. [80] reported values (only in a figure) at the *nominal* temperature of 327°C ($T_r = 1.07$) and densities between 140 and 1000 kg m^{-3}. Bastea and Fried [12] again reported MD-calculated values only in a figure for the *nominal* temperature of 700°C at various pressures, somewhat lower than the NMR values. The discrepancies between the *nominal* temperature and the one inferred from T_r, should be noted; they are due to the low T_c resulting from the water models employed in the MD simulation.

2.4.3 Thermal Conductivity

The thermal conductivity of steam, λ_{th}, including that of SCW, is of interest industrially in connection with power stations and has been measured by several authors over wide temperature and pressure conditions. It has been reported in the proceedings of the International Conferences of the

Properties of Water and Steam over the years. With respect to SCW the name of Franck [81] is connected with the earlier work and that of Tufeu [82–84] with subsequent measurements. The Steam Tables of Haar et al. [2] reported the then available data and Sengers et al. [85] reported representative equations for water, from which the data shown in Table 2.9 are taken.

The thermal conductivity λ_{th} shows a critical enhancement similar to that of the viscosity, persisting up to $T_c + 140K$. As Table 2.9 shows, λ_{th} diminishes to a minimum along both isotherms and isobars, the minimum occurring at increasing temperatures and pressures as these variables are increased. A "background" value $\lambda_{thB} = \sum a_i T^i + \sum b_j \rho^j$ prevails outside the enhancement region, with six a_i temperature coefficients and four b_j density coefficients [84]. To this an "enhancement" $\Delta\lambda_{th}(\rho,T)$ is to be added [82] at $0.4 \leq \rho_r \leq 1.4$ and at $T_r \leq 1.35$ [82]. The enhancement at ρ_c is a function of $\tau = T_r - 1$: $\Delta\lambda_{thc}(\rho_c,\tau)/\lambda_{thB}(\rho_c,T_c) = 0.085\tau^{-0.57}(1 - 0.3\tau^{0.5} - 1.2\tau)$. Sengers et al. [85] gave a 22-parameter smoothing equation for presenting the thermal conductivity over a wide range of thermodynamic states, including SCW and the critical region.

The thermal conductivity of supercritical heavy water was measured by Tufeu et al. [82] in comparison with that of ordinary water. A similar expression for $\lambda_{thB}(D_2O)$ as for $\lambda_B(H_2O)$ but with different coefficients a_i and b_j and the same $\Delta\lambda_{th}(\rho,T)$ as for ordinary water were established. In the supercritical region $\lambda_{th}(H_2O)/\lambda_{th}(D_2O) > 1$ throughout; at 500°C this ratio is 1.03.

TABLE 2.9   The Thermal Conductivity, λ_{th}/mW K^{-1} m^{-1}, of SCW at Some Thermodynamic States (Data from Ref. [85])

	P/MPa						
t/°C	25	30	35	40	50	60	70
375	411.4	438.0	457.5	473.2	498.5	519.4	537.7
400	169.3	330.1	384.5	414.0	451.6	477.7	498.7
450	108.8	136.0	176.5	227.6	315.6	371.2	408.5
500	100.3	113.7	130.7	151.6	202.7	255.6	301.5
550	100.6	109.8	120.8	133.5	163.7	198.0	232.4
600	104.1	111.7	120.3	130.0	152.1	177.0	202.7
650	109.1	115.8	123.3	131.5	149.8	170.1	191.0
700	114.6	120.7	127.5	134.8	150.9	168.5	186.7
750	120.3	126.0	132.5	138.8	153.4	169.1	195.1
800	126.0	131.3	137.0	143.1	156.4	170.6	185.2

2.5   IONIC DISSOCIATION OF SCW

The self-dissociation of water, whether as a liquid or in the supercritical state, is generally written as a bimolecular reaction, because no free protons have been detected in any of the states examined:

$$2H_2O \leftrightarrows H_3O^+ + OH^- \qquad (2.34)$$

Early measurements of the ionic dissociation of neat SCW to 1000°C and 133 kbar were made by means of shock waves by Haman and Linton [86, 87] and by static conductance by Holzapfel [88, 89]. Quist [90] criticized these results on the basis of having probably been affected by impurities from the metallic containers used. He, therefore, preferred to use the conductance of a solution of a hydrolysable salt, NH_4Br, in SCW in comparison with those of KBr and HBr. The equilibria involved are $NH_4Br \leftrightarrows NH_3 + HBr$, $NH_3 + H_2O \leftrightarrows NH_4^+ + OH^-$, and $HBr + H_2O \leftrightarrows H_3O^+ + Br^-$. Quist considered the ionic conductivities of NH_4^+ and K^+ to be the same and included the mean ionic activity coefficients. These were calculated as $y_\pm = \exp(-AI^{1/2}/(1 + I^{1/2}))$, A being the Debye–Hückel coefficient at the prevailing temperature and relative permittivity (Eq. 2.23a) and I the ionic strength. An iterative calculation considering all the species involved, their concentrations and the resulting ionic strength, and the conductivities of the ionic ones yielded the desired ion product constant of SCW, $K_W = [H^+][OH^-]y_\pm^2$, where [] denote molar concentrations. The temperatures ranged up to 800°C, the pressures up to 400 MPa, and the densities were between 450 and 950 kg m^{-3}. The estimated uncertainties were about 0.3–0.5 units in $\log K_W$ for an expression on the molar scale. It involves the (density dependent) molar concentration of water, c_W and is [90]

$$\log K_W = -33.05 - 3050/(T/K) + 16.8 \log c_W \qquad (2.35)$$

Marshall and Franck [63] provided a 7-parameter expression for the entire range of states $0 \leq t/°C \leq 1000$ and $1 \leq P/bar \leq 10{,}000$ based on the publications of many authors. However, for states corresponding to SCW they based their expression on the results of Quist [90]. For the supercritical states of water the expression

$$
\begin{aligned}
-\log[K_W/(\text{mol kg}^{-1})^2] = {} & -24.15 + 0.0259(\rho/\text{kg m}^{-3}) \\
& - 11.56 \times 10^{-6}(\rho/\text{kg m}^{-3})^2 \\
& + (4.058 - 0.001349(\rho/\text{kg m}^{-3})) \\
& + 6.07 \times 10^{-7}(\rho/\text{kg m}^{-3})^2(t/°C)
\end{aligned}
\qquad (2.36)
$$

can be derived. In terms of the temperature and pressure, the following expression holds at ≥ 30 MPa:

$$-\log[K_W/(\text{mol kg}^{-1})^2] = (-15.68 + 0.181(P/\text{MPa}))$$
$$+ (0.0973 - 7.36 \times 10^{-4}(P/\text{MPa}))(t/^\circ\text{C})$$
$$- (5.73 - 0.048(P/\text{MPa})) \times 10^{-5}(t/^\circ\text{C})^2$$
$$(2.37)$$

with an estimated uncertainty of ± 0.03 to ± 0.10. The values of $\log K_W$ of SCW become less negative as the pressure increases but more negative as the temperature increases. Compared to the well known value of $\log[K_W/(\text{mol kg}^{-1})^2] = -14.00$ at 0.1 MPa and 25°C, the values for SCW, except in the critical region, are considerably less negative, increasing to -11.19 at 450°C and 100 MPa and to -8.85 at 600°C and 500 MPa (see Table 2.9). Thus, there can be appreciably more ionic dissociation in SCW than in ambient water. Equation (2.36) was subsequently employed by Marshall [62] for the calculation of the specific conductance of SCW (see Section 2.3.3).

Mesmer et al. [91] obtained the thermodynamic functions ΔH°, ΔS°, ΔC_p°, and ΔV° for the ionization of water from the equilibrium constants. The former two functions at 50 MPa and at 400, 500, 600, and 800°C have the values: $\Delta H^\circ = -197$, -417, -191, and -122 kJ mol^{-1} and $\Delta S^\circ = -520$, -849, -569, and -495 J K^{-1} mol^{-1}. The ionization is thus entropy controlled. They also used the volume change ΔV° during ionic dissociation of water in SCW to estimate that about 14 water molecules are associated with the ions

TABLE 2.10   The p(K_W/mol^2 kg^{-2}) (Molal Scale) of SCW According to Data from Ref. [63] and Eq. (2.37)

$t/^\circ$C	P/MPa						
	25	30	35	40	50	60	70
400	13.16	12.97	12.79	12.61	12.24	11.88	11.51
450	15.18	14.91	14.65	14.38	13.85	13.32	12.79
500	16.97	16.64	16.30	15.97	15.30	14.63	13.96
550	18.54	18.14	17.75	17.36	16.57	15.79	15.00
600	19.88	19.44	19.00	18.56	17.68	16.80	15.93
650	20.99	20.52	20.04	19.57	18.63	17.68	16.73
700	21.88	21.38	20.89	20.39	19.40	18.41	17.42
750	22.54	22.03	21.53	21.02	20.01	19.00	17.99
800	22.97	22.47	21.97	21.46	20.45	19.45	18.45

formed at a density of $0.8\,g\,cm^{-3}$, a value that increases with decreasing densities. Chen et al. [118] measured calorimetrically the enthalpy of ionization of water along the saturation lineup to 350°C. They derived $\Delta H°$ and $\Delta S°$ values that could be extrapolated to the SCW conditions and yielded pK_W values in good agreement with those of Quist [90] and of Marshall and Franck [63].

Tanger and Pitzer [92] applied to SCW a different approach, not depending on the conductance of electrolytes. Its results agreed with those of Marshall and Franck [63] at $\rho \geq 450\,kg\,m^{-3}$ but was said to produce more correct values for SCW at lower densities, and was applicable up to 2000°C and 500 MPa. It was based on a thermodynamic cycle involving estimates of the Gibbs energies of hydration of H_3O^+ and OH^- as well as the well established values for the vaporization of water (not relevant for SCW), and its dissociation in the vapor phase.

$$2H_2O(g, T) \qquad \rightarrow H_3O^+(g, T) \qquad + OH^-(g, T)$$
$$\updownarrow \qquad\qquad\qquad \updownarrow \qquad\qquad\qquad \updownarrow \qquad\qquad (2.38)$$
$$2H_2O(SCW, P, T) \quad \rightarrow H_3O^+(SCW, P, T) \quad + OH^-(SCW, P, T)$$

The Gibbs energies of hydration were estimated according to a semicontinuum model, involving the enthalpy and entropy changes of successive hydration of the ions. The parameters of the model were fitted, however, to the experimental K_W values of Quist [90] and Sweeton et al. [93] along the saturation curve in order to obtain the best fit to SCW. Tawa and Pratt [94] applied a theoretical calculation to the ionic product of water at SCW conditions up to 600°C (and also to thermodynamic states of liquid water), based on a dielectric continuum model. It assumes a cavity of radius R, within which the relative permittivity is fixed at 2.42, being the same for the reactants and products of Eq. (2.33), but depending on the temperature and density. At a density of $800\,kg\,m^{-3}R$ increases from $0.184\,nm$ at 400°C to $0.192\,nm$ at 600°C, and the density dependence, between 600 and $1050\,kg\,m^{-3}$ at a constant temperature, is concave downward but does not exceed $0.002\,nm$.

Bandura and Lvov [95] developed a semitheoretical approach to fit the available experimental data for K_W over wide ranges of density and temperature. The final formulation has a simple analytical form, includes seven adjustable parameters, and fitted, within experimental uncertainties, the available K_W data up to 800°C and up to $1200\,kg \cdot m^{-3}$. Takahashi et al. [96] applied a quantum mechanical/molecular mechanical method for the estimation of the ionic dissociation of SCW. However, they used the fixed temperature 600K, definitely below the experimental critical temperature for water, but it corresponds to $T_r = 1.07$ for the TIP4P water model employed. The relevance of their results for SCW is therefore questionable.

Halsted and Masters [97] applied molecular dynamics simulations with the SPC/E model of water to obtain the ion product of water at a fixed pressure of 32 MPa and temperatures ranging from ambient to 427°C, that is, into the SCW range. The simulations reproduced the minimum in pK_W near 220°C, but the values for SCW depended on the number of water molecules considered in the simulations. The minimum in pK_W is interpreted in terms of the endothermic autoionization process, tending to increase the ion concentration with increasing temperatures on the one hand and the decreasing permittivity of SCW disfavoring the formation of ions on the other hand.

2.6   PROPERTIES RELATED TO THE SOLVENT POWER OF SCW

The Hildebrand solubility parameter, δ_H, of SCW is of possible use in the estimation of the solubilities of various substances in it. Obviously, the usual expression employed in obtaining solubility parameters, $\delta_H = [(\Delta_V H - RT)/V]^{1/2}$, is not applicable in supercritical fluids because the enthalpy of vaporization is meaningless for them. However, the quantity $\Delta_V H - RT$ actually stands for the configurational potential energy $-U$ of the supercritical solvent. Hence, the appropriate expression to be employed is $\delta_H = [-U/V]^{1/2}$. The Redlich–Kwong EoS is known to be applicable to SCFs [98]:

$$P = RT/(V - b) - N_A^2 \alpha a/V(V + b) \qquad (2.39)$$

Here b and a are parameters descriptive of the excluded volume and attractive forces, respectively, and $\alpha = T^{-1/2}$. For SCW modifications of the Redlich–Kwong EoS, for example, those of Soave–Redlich–Kwong (SRK) [99] or of Peng–Robinson (PR) [100] are required, because water is not a conformal fluid interacting by dispersion forces only. These modifications essentially replace the $\alpha = T^{-1/2}$ factor multiplying the attractive potential a in Eq. (2.39) by $\alpha = [1 + m(1 - T_r^{1/2})]^2$, where m is a quadratic in ω, the Pitzer acentric factor that is 0.3443 for water. The quadratics for $m(\omega)$ given by SRK and by PR differ

$$m(\text{SRK}) = 0.480 + 1.574\omega - 0.176\omega^2 \qquad (2.40a)$$

$$m(\text{PR}) = 0.37464 + 1.54226\omega - 0.26992\omega^2 \qquad (2.40b)$$

The ratio of the two specifications of m for water is 1.145. Furthermore, these two approaches specified the attractive parameter $a = cR^2 T_c^2/P_c$ with somewhat different values of c: $c(\text{SRK}) = 0.42748$ and $c(\text{PR}) = 0.45724$, reducing slightly the differences between the resulting αa values. Finally, the configurational potential energy is simply related to the attractive potential parameter of the EoS: $-U = a\alpha/V$. The expression for the Hildebrand solubility parameter of nonconformal fluids is then [101]

$$\delta_H/MPa^{1/2} = 2.95\,(c/0.4278)^{1/2}(P_c/MPa)^{1/2}\,[1 + m(1 - T_r^{1/2})]T_r^{1/4}\rho_r$$

(2.41)

At a given (reduced) density, the factor in the square brackets reduces the value of δ_H as the (reduced) temperature increases more than $T_r^{1/4}$ increases it. The differences between the SRK- and PR-derived values of $\delta/MPa^{1/2}$ differ by only 0.1–0.2 units, so that their averages may be used. The resulting values for SCW are

$$\delta_H(SCW)/MPa^{1/2} = 14.4[1 - 3.6(T_r - 1)]\rho_r \qquad (2.42)$$

These values are much below that for liquid water at ambient conditions, 48 $MPa^{1/2}$ at all practically available thermodynamic states of T and ρ.

The total Hildebrand solubility parameter of SCW, δ_H as presented by Eq. (2.42) may not be the best indicator for the solubility of specific solutes. Its division into the three Hansen solubility parameters: δ_d for dispersion interactions, δ_p for polar interactions, and δ_h for hydrogen bonding could solve this problem.

$$\delta_H^2 = \delta_d^2 + \delta_p^2 + \delta_h^2 \qquad (2.43)$$

For water at 25°C these reference values are 15.6, 16.0, and 42.3 $MPa^{1/2}$, respectively. Williams et al. [102] provided expressions for calculating such values at different temperatures and pressures, including those pertinent to supercritical conditions, albeit specifically for supercritical carbon dioxide. Still, the relevant expressions are general

$$\delta_d = \delta_{d\,ref}(V/V_{ref})^{1.25} \qquad (2.44a)$$

$$\delta_p = \delta_{p\,ref}\,(V/V_{ref})^{0.5} \qquad (2.44b)$$

$$\delta_h = \delta_{h\,ref}\,\exp\{0.00132(T - T_{ref}) + 0.5\ln(V/V_{ref})\} \qquad (2.44c)$$

with $V_{ref} = 18.07\,cm^3\,mol^{-1}$ at $T_{ref} = 298.15K$. For SCW the molar volume $V/cm^3\,mol^{-1}$ at the desired temperature T and pressure should be obtained from the Steam Tables, EoSs, or Table 2.1 for a limited set of conditions. A figure showing δ_h for $380 \leq t/^\circ C \leq 480$ and $15 \leq P/MPa \leq 40$ was shown by Morimoto et al. [103]. Over these temperature and pressure ranges they increase with pressure at isotherms (from 6 to $16\,MPa^{1/2}$ at $420^\circ C$) and diminish with rising temperatures at isobars (from 17 to $7\,MPa^{1/2}$ at 25 MPa). Very recently Marcus [119] reported the individual Hansen parameters δ_d, δ_p, and δ_h for SCW at $400 \leq t/^\circ C \leq 700$ and $25 \leq P/MPa \leq 100$ and discussed briefly their applications.

Solvatochromic probes have found extensive use for the estimation of properties, such as solvation ability, polarity, and hydrogen bonding ability, of solvents at ambient conditions. Little use has been made of otherwise suitable probes in SCW, because they tend to be unstable and if air is not carefully excluded they are oxidized, with oxidation products that have similar but not exactly the same solvatochromic properties. In any case, only relatively low temperatures could be used, $T_r \leq 1.2$. Lu et al. [104, 105] studied the application of the Kamlet–Taft solvatochromic probes in near critical water, along the saturation curve up to $275^\circ C$, but not to SCW.

Bennet and Johnston [106] were the first who actually applied UV-visible spectroscopy to π^* (polarity/polarizability) probes in SCW up to $440^\circ C$, namely benzophenone and acetone, since the generally employed nitroaromatics were unstable in SCW. Oka and Kajimoto [107] and Minami et al. [108] measured the UV spectrum of 4-nitroanisole in SCW in a flow system in order to minimize the decomposition and calculated the Kamlet–Taft solvatochromic parameter π^*. A linear dependence of π^* on the density was found for SCW at 400 and $420^\circ C$:

$$\pi^* = -(0.71 \pm 0.06) + (1.77 \pm 0.09)\rho_w \qquad (2.45)$$

A more comprehensive account of the applicability of solvatochromic probes in SCW is presented in Chapter 4 (Section 4.2.2).

2.7 SUMMARY

Supercritical water, being a single component occupying a single phase, has two degrees of freedom according to the phase rule (1.1). Of the three coupled variables: the temperature T, the pressure P, and the density ρ (or the volume of a given amount of substance V) two can be chosen at will, determining thereby the third. This property, the ability to tune the density by choosing proper temperature and pressure is of paramount importance for applications of SCW, the properties of which depend strongly on its density. Hence, knowledge of the PVT data of SCW is a prerequisite for its use. This information is available in extensive tables (Section 2.1.1), of which Table 2.1 is a very abbreviated one. Else, an equation of state from among the several that have been proposed

(Sections 2.1.2 and 2.1.3) may be invoked in order to calculate the compressibility factor $Z = PV/RT$, Eqs. (2.1)–(2.3), (2.6), and (2.12), that relates the P–V–T values to one another. The former three expressions are empirical fitting ones with as many as 15 parameters (Table 2.2), the latter expressions for Z as well as other ones (not shown) derived from theoretically based EoSs, may have fewer parameters. These EoSs take the hydrogen bonding existing in SCW into account, hence yield a measure of the extent of hydrogen bonding (f_0, the fraction of *non*hydrogen bonded water molecules, fully discussed in Section 3.4). The thermal expansibility and the compressibility of SCW are immediate results from the PVT relationships.

Certain thermodynamic quantities can be derived from the EoSs, foremost among them is the heat capacity, both the isochoric (constant volume) and isobaric (constant pressure) one. Having diverged to infinity, the heat capacity diminishes to more usual values at several degrees beyond T_c. A brief survey of the data is shown in Table 2.3. At given pressures the molar heat capacity C_p has a maximum at increasing temperatures and then decreases slowly up to very high temperatures. At given temperatures the pressure dependence C_p again shows maxima at moderate to high pressures that become shallower at increasing temperatures. The isochoric heat capacity of SCW, most definitely studied by Abdulagatov et al. [13] and subsequently, is related to the heat capacity of ideal gas water by Eqs. (2.15) and (2.12), the latter describing an EoS.

The molar enthalpy and entropy of SCW, derived from the EoSs, are shown in Table 2.4. Expressions for the residual quantities, that is the values after subtraction of the ideal gas values, are described by Eqs. (2.17) and (2.18). The residual Helmholtz energy and the fugacity ratio are also dealt with briefly in Section 2.2.2. The ultrasound velocity in SCW, represented by Eq. (2.19) permits the calculation of the adiabatic compressibility when combined with the density data (Table 2.1).

Of the electrical and optical properties of SCW the most important one is the relative permittivity, ε_r, the values of which were most definitely established by Fernandez et al. [57] up to 600°C and to 1.2 GPa (Table 2.5). They based their calculation on a modification of the Kirkwood–Fröhlich expression involving the dipole orientation parameter g, namely Eq. (2.22). The light refraction by SCW has been reported up to 500°C and 100 MPa over a wide wavelength range (Table 2.7). Its square at infinite frequency (approximated as that at 200 nm wavelength), n_{200}^2, is used to represent the atomic/electronic response of the water molecules in an electric field, in Eq. (2.22). The quantity g was then modeled by means of a 14-parameter expression in the reduced temperature and density, which permitted the estimation of derivative functions with respect to the temperature and the pressure as well as the limiting slopes of the Debye–Hückel expression for the osmotic and activity coefficients of electrolyte solutions, Eq. (2.23). The values of ε_r diminish with increasing temperatures and increase with raised pressures. At the lower temperature they are of the order of moderately polar solvents but diminish toward those of nonpolar solvents at $\geq 600°C$. The permittivity of SCW as a function of the frequency of the electrical field was studied by microwave spectroscopy, but the derived dielectric relaxation time τ_D, Eq. (2.24), is more fully discussed in Section 3.3.4.

The specific electrical conductance of pure water at ambient conditions is very small, but it increases considerably as the temperature and pressure are increased and reaches appreciable values in SCW at not too low densities. The relevant data were obtained by Marshall [62] at up to $1000°C$ and $1000\,MPa$ and are shown in Table 2.6. The data are related to the ionic dissociation of SCW dealt with in Section 2.7.

The bulk transport properties of SCW include its dynamic viscosity, thermal conductivity, and self-diffusion. The dynamic viscosity, η, shown in Table 2.8, can be summarized by a 5-parameter equation in the temperature and density, Eq. (2.30). It is valid for $400 \leq t/°C \leq 800$ and pressures of $25 \leq P/MPa \leq 100$ (densities $50 \leq \rho/kg\,m^{-3} \leq 800$. At low, gas like, densities of SCW its viscosity increases with increasing temperatures as expected from the Enskog theory, but for isochors at high densities, where SCW exhibits liquid-like properties, the viscosity still increases with increasing temperatures, contrary to expectations. The self-diffusion coefficient, D, of SCW was measured by the NMR method for $400 \leq t/°C \leq 700$ with $22 \leq P/MPa \leq 145$. At a density of $400\,kg\,m^{-3}$ the values of $D/10^{-9}\,m^2\,s^{-1}$ were 90 at $400°C$ increasing to 108 at $700°C$ and increased with diminishing densities. The results could be fitted by Eq. (2.31). The thermal conductivity of SCW, λ_{th}, is of interest industrially in connection with power stations and nuclear reactors. The Steam Tables [2] reported the then available data shown in Table 2.9. The dynamic viscosity η and the thermal conductivity λ_{th} show a critical enhancement persisting to tens of K beyond T_c.

The self-dissociation of SCW is generally written as a bimolecular reaction, because no free protons have been detected in any of the states examined: $2H_2O \leftrightharpoons H_3O^+ + OH^-$. The ion product constant of SCW, K_W, for temperatures ranging up to $800°C$, pressures up to $400\,MPa$, and densities between 450 and $950\,kg\,m^{-3}$ yields log K_W on the molar scale involving the (density dependent) molar concentration of water, c_W, Eq. (2.34). The 6-parameter expressions (2.35) in terms of density and (2.36) in terms of pressure, in addition to the temperature, hold at $P \geq 30\,MPa$. Values of $p(K_W/mol^2\,kg^{-2})$ (molal scale) for SCW are shown in Table 2.10. It is noteworthy that pK_W has a minimum at $220°C$, and although the ionization diminishes in SCW compared with hot compressed water, the acidity of SCW (hydrogen ion concentration) is larger than what is commonly expected for water, given its $pK_W = 14$ at ambient conditions.

Some of the bulk properties of SCW are relevant to its properties as a solvent, fully discussed in Chapter 4. One such property is its Hildebrand solubility parameter, δ_H, given by Eq. (2.42) and seen to be proportional to the reduced density, ρ_r and only mildly dependent on the temperature. At SCW states near the critical $\delta_H(SCW)/MPa^{1/2} \sim 14$, considerably less than for ambient conditions, $\delta_H/MPa^{1/2} \sim 48$. This then leads to good solubilities of solutes with $10 \leq \delta_H/MPa^{1/2} \leq 18$, according to the general rule for solubilities allowed by the solubility parameter approach. However, the solubilities of nonpolar solutes with $\delta_H/MPa^{1/2} \ll 10$, is not explained by the derived values for SCW. The Hansen solubility parameters, Eq. (2.44a), recently [119] available for SCW, might help in this respect. Solvatochromic parameters for SCW are not generally available, because the commonly used indicators are unstable in SCW. Still, values of the polarity/polarizability parameter π^* are available for SCW up to 420°C, Eq. (2.45).

It should be noted that many of the bulk properties of SCW briefly described above are also known for supercritical heavy water, D_2O, as reported in the appropriate places. Also of note is the fact that many of the experimentally measured properties have also been obtained by computer simulations with appropriate potential models for the water molecule. These simulation results are also dealt with as required.

REFERENCES

1. E. Schmidt, in U. Grigull, Ed., *Properties of Water and Steam in SI-Units, 2nd revised printing*, Springer, Berlin, 1979.

2. L. Haar, J. S. Gallagher, and G. S. Kell, *NBS/NRC Steam Tables*, Hemisphere Publ., Washington, 1984. Data to 2000°C and 30,000 bar (for C_p).

3. U. Grigull, J. Straub, and P. Schiebener,Eds., *Steam Tables in SI-Units*, Springer, Heidelberg, 1990. Data to 800°C and 1000 bar.

4. N. E. Dorsey *Properties of Ordinary Water Substance*, Hafner Publ. Co., New York, 1968.

5. A. Saul and W. Wagner, *J. Phys. Chem. Ref. Data* **18** 1537 (1989).

6. W. Wagner and A. Pruss, *J. Phys. Chem. Ref. Data* **31**, 387 (2002).

7. Z. Duan, N. Møller, and J. H. Weare, *Geochim. Cosmochim. Acta* **56**, 2605 (1992).

8. D. M. Kerrick and G. K. Jacobs, *Am. J. Sci.* **281**, 735 (1981).

9. M. C. Kutney, V. S. Dodd, K. A. Smith, H. J. Herzog, and J. W. Tester *Fluid Phase Equilib.* **128**, 149 (1997).

10. S.-M. Chern, Proceedings of the National Spring Meeting AIChE 2007, p81312/1, 2007.

11. E. H. Abramson and J. M. Brown, *Geochim. Cosmochim. Acta* **68**, 1827 (2004).

12. S. Bastea and L. E. Fried, *J. Chem. Phys.* **128**, 174502 (2008).

13. I. M. Abdulagatov, V. I. Dvoryanchikov, and A. N. Kamalov, *J. Chem. Eng. Data* **43**, 830 (1998).

14. P. J. Smits, I. G. Economou, C. J. Peters, and J. de S. Arons *J. Phys. Chem.* **98**, 12080 (1994).

15. R. B. Gupta, C. G. Panayiotou, I. C. Sanchez, and K. P. Johnston, *AIChE J.* **38**, 1243 (1992).

16. C. Panayiotou and I. C. Sanchez, *J. Phys. Chem.* **95**, 10090 (1991).

17. R. B. Gupta and K. P. Johnston, *Fluid Phase Equilib.* **99**, 133 (1994).

18. W. G. Capman, K. E. Gubbins, G. Jackson, and M. Radosz, *Ind. Eng. Chem. Res.* **29**, 1709 (1990).

19. S. H. Huang and M. Radosz, *Ind. Eng. Chem. Res.* **29**, 2284 (1990).

20. G. D. Ikonomou and M. D. Donohue, *AIChE J.* **32**, 1716 (1986).

21. I. G. Economou and M. D. Donohue, *Ind. Eng. Chem. Res.* **31**, 2388 (1992).

22. H. Touba and G. A. Mansoori, *Fluid Phase Equilib.* **150–151** 459 (1998).

23. T. Vlachou, I. Prinos, J. H. Vera, and C. G. Panayiotou, *Ind. Eng. Chem. Res.* **41**, 1057 (2002).

24. Z.-Q. Hu, J.-C. Yang, and Y.-G. Li, *Fluid Phase Equilib.* **205**, 1 (2003).

25. M. D. Bermejo, A. Martin, and M. J. Cocero, *J. Supercrit. Fluids* **42**, 27 (2007).

26. A. Anderko and K. S. Pitzer, *Geochim. Cosmochim. Acta* **57**, 1657 (1993).

27. J. J. Kosinski and A. Anderko, *Fluid Phase Equilib.* **183–184** 75 (2001).

28. S. B. Kiselev, I. M. Abdulagatov, and A. H. Harvey, *Int. J. Thermophys.* **20**, 563 (1999).

29. N. V. Tsederberg, A. A. Aleksandrov, T. S. Khasanshin, and D. K. Larkin, *Teploenergetika* **20**, 13 (1973).

30. W. L. Marshall and J. W. Simonson, *J. Chem. Thermodyn.* **23**, 613 (1991).

31. A. R. Bazaev, I. M. Abdulagatov, J. W. Magee, E. A. Bazaev, and A. E. Ramazanova, *J. Supercrit. Fluids* **26**, 115 (2003).

32. H. W. Woolley, *J. Res. Natl. Bur. Std.* **92**, 35 (1987).

33. J. R. Cooper, *Int. J. Thermophys.* **3**, 35 (1983).

34. N. B. Vargaftig, *Handbook of Physical Properties of Liquids and Gases* Hemisphere, New York, 1983.

35. A. M. Sirota and P. E. Belyakova, *Teploenergetika* **13**, 84 (1966).

36. Y. Marcus *J. Solution Chem.* **25**, 455 (1996).

37. A. M. Sirota and B. K. Maltsev, *Teploenergetika* **9** 52, 70 (1962).

38. J. M. H. Levelt Sengers, B. Kamgar-Paral, F. W. Balfour, and J. V. Sengers, *J. Phys. Chem. Ref. Data* **12**, 1 (1983).

39. B. A. Mursalov, I. M. Abdulagatov, V. I. Dvoryanchikov, A. N. Kumalov, and S. B. Kiselev, *Int. J. Thermophys.* **20**, 1497 (1999).

40. I. M. Abdulagatov, J. W. Magee, S. B. Kiselev, and D. G. Friend,in P. R. Tremaine, Ed., Steam, Water, and Hydrothermal Systems, Proceedings of the 13th International Conference on Properties of Water and Steam, 1999 Toronto, Ont., Canada, 2000, p. 374.

41. A. I. Abdulagatov, G. V. Stepanov, I. M. Abdulagatov, A. E. Ramazanova, and G. S. Alisultanova, *Chem. Eng. Commun.* **190**, 1459 (2002).

42. A. Abdulagatov, G. V. Stepanov, and A. Abdulagatov, *Chem. Eng. Commun.* **190**, 1499 (2003).

43. S. L. Rivkin and B. N. Egorov, *Teploenergetika* **9** 60 (1962);S. L. Rivkin and B. N. Egorov, *Teploenergetika* **10**, 75 (1963).

44. N. G. Polikhronidi, I. M. Abdulagatov, J. W. Magee, and G. V. Stepanov *Int. J. Thermophys.* **24**, 405 (2003).

45. E. Schmidt, *Landoldt-Börnstein*, 6th ed., Vol. IV/4a, Springer, Berlin, 1967.

46. H.-J. Kretzschmar, J. R. Cooper, J. S. Gallagher, et al., *J. Eng. Gas Turb. Power* **129**, 294, 1125 (2007).

47. J. P. Petitet, L. Denielou, R. Tufeu, and B. Le Neindre, *Int. J. Thermophys.* **7**, 1065 (1986).

48. E. U. Franck, *Z. Phys. Chem. (NF)* **8**, 107 (1956).

49. A. S. Quist and W. L. Marshall, *J. Phys. Chem.* **69**, 3165 (1965).

50. Yu. M. Lukashov, B. P. Golubev, and F. B. Ripol-Saragosi, *Teploenergetika* **22**, 79 (1975).

51. K. Heger, M. Uematsu, and E. U. Franck, *Ber. Bunsenges. Phys. Chem.* **84**, 758 (1980).

52. M. Uematsu and E. U. Franck, *J. Phys. Chem. Ref. Data* **9**, 1291 (1980).

53. D. G. Archer and P. Wang, *J. Phys. Chem. Ref. Data* **19** 371 (1990).

54. Yu. V. Mulev and S. N. Smirnov, *Teplofizika Vysokikh Temp.* **30**, 51 (1990).

55. E. U. Franck, S. Rosenzweig, and M. Christoforakos, *Ber. Bunsenges. Phys. Chem.* **94**, 199 (1990).

56. K. S. Pitzer, *Proc. Natl. Acad. Sci. USA* **80**, 4575 (1983).

57. D. P. Fernandez, A. R. H. Goodwin, E. W. Lemmon, J. M. H. Levelt Sengers, and R. C. Williams, *J. Phys. Chem. Ref. Data* **26**, 1125 (1997).

58. Y. Marcus, *Fluid Phase Equilib.* **164**, 131 (1999).

59. E. Wasserman, B. Wood, and J. Brodholt, *Ber. Bunsenges. Phys. Chem.* **98**, 906–911 (1994).

60. E. Wasserman, B. Wood, and J. Brodholt, *Cosmochim. Geochim. Acta* **59**, 1 (1995).

61. K. Okada, Y. Imashuku, and M. Yao, *J. Chem. Phys.* **107**, 9302 (1997).

62. W. L. Marshall, *J. Chem. Eng. Data* **32**, 221 (1987).

63. W. L. Marshall and E. U. Franck, *J. Phys. Chem. Ref. Data* **10**, 295 (1981).

64. P. Schiebener, J. Straub, J. M. H. Levelt Sengers, and J. S. Gallagher, *J. Phys. Chem. Ref. Data* **19**, 677 (1990).

65. K. H. Dudziak and E. U. Franck, *Ber. Bunsenges. Phys. Chem.* **70**, 1120 (1966).

66. J. T. R. Watson, R. S. Basu, and J. V. Sengers, *J. Phys. Chem. Ref. Data* **9**, 1255 (1980).

67. J. V. Sengers and B. Kamgar-Paral, *J. Phys. Chem. Ref. Data* **13**, 185 (1984).

68. M. J. Assael, E. Bekou, D. Glakounakis, D. G. Friend, M. A. Kileen, J. Millat, and A. Nagashima, *J. Phys. Chem. Ref. Data* **23**, 141 (2000).

69. O. Hirschfelder, C. F. Curtiss, and R. B. Bird, *Theory of Gases and Liquids*, Wiley, New York, 1954.

70. S. L. Rivkin, A. Ya. Levin, L. B. Izrailevskii, K. G. Kharitonov, and V. N. Ptitsnaya, *Teploenergetika*, 65 (1977).

71. W. J. Lamb, G. A. Hoffman, and J. Jonas *J. Chem. Phys.* **74**, 6875 (1981).

72. E. Wilhelm, *J. Chem. Phys.* **58**, 3558 (1973).

73. K. Yoshida, Ch. Wakai, N. Matubayashi, and M. Nakahara, *J. Chem. Phys.* **123**, 164506–1 (2005).

74. K. Yoshida, N. Matibayasi, Y. Uosaki, and M. Nakahara, *J. Phys. Conf. Ser.*, **215**, 012093 (2010).

75. I. A. Beta, J.-C. Li, and M.-C. Bellissent-Funel, *Chem. Phys.* **292**, 229 (2003).

76. T. Tassaing, Y. Danten, and M. Besnard, *Pure Appl. Chem.* **76**, 133 (2004).

77. Q.-Y. Tong, G.-H. Gao, M.-H. Han, and Y.-X. Yu, *Int. J. Thermophys.* **23**, 635 (2002).

78. A. G. Kalinichev, *Ber. Bunsenges. Phys. Chem.* **97**, 872 (1993).

79. T. I. Mizan, P. E. Savage, and R. M. Ziff, *J. Phys. Chem.* **98**, 13067 (1994).

80. N. Yoshii, H. Yoshie, S. Miura, and S. Okazaki, *J. Chem. Phys.*, **100**, 4873 (1998).

81. E. U. Franck, *Ber. Bunsenges. Phys. Chem.* **88**, 820 (1984).

82. R. Tufeu, P. Bury, Y. Garrabos, and B. Le Neindre, *Int. J. Thermophys.* **7**, 663 (1986).

83. R. Tufeu and B. Le Neindre, *Int. J. Thermophys.* **8**, 283 (1987).

84. Ph. Desmarest and R. Tufeu, *Int. J. Thermophys.* **11**, 1035 (1990).

85. J. V. Sengers, J. T. R. Watson, R. S. Basu, B. Kamgar-Paral, and R. C. Hendricks, *J. Phys. Chem. Ref. Data* **13**, 893 (1984).

86. S. D. Haman and M. Linton, *Trans. Faraday Soc.* **62**, 2284.

87. S. D. Haman and M. Linton, *Trans. Faraday Soc.* **65**, 2186 (1969).

88. W. Holzapfel and E. U. Franck, *Ber. Bunsenges. Phys. Chem.* **70**, 1105 (1966).

89. W. Holzapfel, *J. Chem. Phys.* **50**, 4424 (1969).

90. A. S. Quist, *J. Phys. Chem.* **74**, 3396 (1970).

91. R. E. Mesmer, W. L. Marshall, D. A. Palmer, J. M. Simonson, and H. F. Holmes, *J. Solution Chem.* **17**, 699 (1988).

92. J. C. Tanger, IV, and K. S. Pitzer, *AIChE J.*, **35**, 1631 (1989).

93. F. H. Sweeton, R. E. Mesmer, and C. G. F. Baes, Jr., *J. Solution Chem.* **3**, 191 (1974).

94. G. J. Tawa and L. R. Pratt, *J. Am. Chem. Soc.* **117**, 1625 (1995).

95. A. V. Bandura and S. N. Lvov, *J. Phys. Chem. Ref. Data* **35**, 15 (2006).

96. H. Takahashi, W. Satou, T. Hori, and T. Nitta, *J. Chem. Phys.* **122**, 044504 (2005).

97. S. J. Halstead and A. J. Masters, *Mol. Phys.* **108**, 193 (2010).

98. S. Goldman, C. G. Gray, W. Li, B. Tomberli, C. G. Joslin, *J. Phys. Chem.* **100**, 7246 (1996).

99. G. Soave, *Chem. Eng. Sci.* **27**, 1197 (1972).

100. D. Y. Peng and D. B. A. Robinson, *Ind. Eng. Chem., Fundam.* **15**, 59 (1976).

101. Y. Marcus, *J. Supercrit. Fluids* **38**, 7 (2006).

102. L. L. Williams, J. B. Rubin, and H. W. Edwards, *Ind. Eng. Chem. Res.* **43**, 4967 (2004).

103. M. Morimoto, S. Sato, and T. Takanohashi, *J. Jpn. Petrol. Inst.* **53**, 61 (2010).

104. J. Lu, J. S. Brown, C. L. Liotta, and C. A. Eckert, *Chem. Commun.* 665 (2001).

105. J. Lu, J. S. Brown, E. C. Boughner, C. L. Liotta, and C. A. Eckert, *Ind. Eng. Chem. Res.* **11**, 2835 (2002).

106. G. E. Bennett and K. P. Johnston, *J. Phys. Chem.* **98**, 441 (1994).

107. H. Oka and O. Kajimoto, *Phys. Chem. Chem. Phys.* **5**, 2535 (2003).

108. K. Minami, K. Ohashi, M. Suzuki, T. Aizawa, T. Adschiri, and K. Arai, *Anal. Sci.* **22**, 1417 (2006).

109. M. S. Skaf and D. Laria, *J. Chem. Phys.* **113**, 3499 (2000).

110. E. Guardia and J. Marti, *J. Mol. Liquids* **101**, 137 (2002).

111. M. Yao and Y. Hiejima, *J. Mol. Liquids* **96–97**, 207 (2002).

112. E. Gardia and J. Marti, *Phys. Rev. E* **69**, 011502–1 (2004).

113. A. M. Sirota and A. J. Grishkov, *Teploenergetika* **13**, 61 (1966).

114. Y. Marcus, *J. Mol. Liq.* **81**, 101 (1999).

115. K. Okada, M. Yao, Y. Hiejima, H. Kohno, and Y. Kajihara, *J. Chem. Phys.* **110**, 3026 (1999).

116. M. A. Anisimov, S. B. Kiselev, and I. G. Kostyukova, *Teplofiz. Vysok. Temp.* **25**, 31 (1987).

117. K. Kuge, Y. Murayama, T. Honda, Y. Kato, and Y. Yoshizawa, *Progr. Nucl. Energy* **37**, 441 (2000).

118. X. Chen, J. L. Oscarson, S. E. Gillespie, H. Cao, and R. M. Izatt, *J. Sollution Chem.* **23**, 747 (1994).

119. Y. Marcus, *J. Supercrit. Fluids* **62**, 60 (2012).

3

MOLECULAR PROPERTIES OF SCW

The molecular properties of supercritical water (SCW) are reflected by its structure and dynamics. The structure is governed by the hydrogen bonding between the water molecules, if any, and the repulsion at short distances and dispersion and dipole interactions between neighboring molecules. The dynamics are mainly expressed in terms of the lifetime of a hydrogen bond in particular and the orientation relaxation times of water molecules. Both the structure and dynamics can be accessed by experimental methods, mostly diffraction of X-rays and neutrons for the structure and spectroscopy, such as NMR, dielectric relaxation, and ultrafast vibrational spectroscopy. Computer simulations, Monte Carlo (MC) and molecular dynamics (MD) for the structure and the latter only for the dynamics, as well as theoretical studies complement the information from the experimental investigations.

One way of looking at the molecular properties of SCW is to consider how these affect its structure. The question of what is meant by the structure of a fluid arises in this connection. This was answered by Gorbaty and Gupta [1] as the variety of preferential molecular arrangements due to (thermal) fluctuations in the nearest environment of a given molecule.

The molecular arrangements are described by the radial distribution function of the fluid, $N(r)$. The differential of the radial distribution function of a fluid, dN, is the probability of finding a particle in a spherical shell of thickness dr at a distance r from a given particle. At large distances, there are no interactions between the particles so that $dN(r \rightarrow \infty, dr) = \rho 4\pi r^2 dr$, and is proportional to the number density of particles, ρ. At short distances attraction and repulsion forces between the particles affect the probability for another

Supercritical Water A Green Solvent: Properties and Uses, First Edition. Yizhak Marcus.
© 2012 John Wiley & Sons, Inc. Published 2012 by John Wiley & Sons, Inc.

particle to be in a volume element at the distance r from a given one: they are correlated. Hence the pair correlation function is defined by the conditional probability of finding a particle at a distance r from another particle:

$$dN(r, dr) = g(r)\rho 4\pi r^2 dr \tag{3.1}$$

No correlation exists between particles at large distances from each other, so that $g(r \rightarrow \infty) = 1$. On the other hand, at distances shorter than the diameter σ of the particles the large repulsion of the electronic shells of the atoms prevent their overlapping and $g(r \leq \sigma) = 0$. Integration of Eq. (3.1) from σ to any distance r yields the number of neighboring molecules around the particle at the origin, that is, the coordination number N_{co} up to that distance. Generally such coordination numbers reach plateaus (have small slopes) at $r \sim \sigma$, that is the first coordination shell, and possibly also the second at $r \sim 2\sigma$).

The molecular pair correlation functions $g(r)$ of SCW at various thermodynamic states provide some of the desired information on the structure of SCW. However, these quantities are integrated over the orientations of the molecules, so that some information is still lacking. The partial pair correlation functions pertaining to specific pairs of atoms: $g(O-O, r)$, $g(O-H, r)$, and $g(H-H, r)$ do provide the required information.

The distinguishing feature of water from many other fluids is the hydrogen bonding that takes place between pairs of molecules. There is sufficient evidence now that hydrogen bonds do exist in SCW, with general agreement that the tetrahedral hydrogen-bonded network present in ambient water is no longer present in SCW. There is also general agreement among investigators that the extent of hydrogen bonding in SCW increases with increasing densities but diminishes with increasing temperature. The extent of this hydrogen bonding in SCW, whether it consists of dimers only or whether larger clusters: trimers, tetramers, and so on also exist, and the thermodynamic states of SCW at which such species can be found are discussed in this chapter. Information on this is obtained from diffraction measurements, computer simulations, spectroscopic data, and from analyzing the findings on bulk properties dealt with in Chapter 2.

A stumbling block in the path to elucidation of such problems is the lack of a precise definition of the criteria for an existing, intact hydrogen bond or for its complete breaking. Investigators using *ab initio* theoretical computations and molecular dynamics computer simulations of water at ambient conditions considered the following as criteria for the existence of a hydrogen bond between two water molecules, as suggested by Kumar et al. [2] (Fig. 3.1):

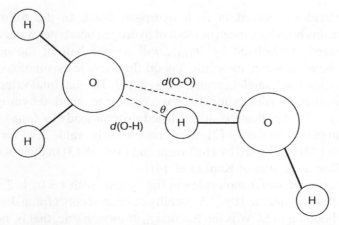

FIGURE 3.1 Schematic representation of a hydrogen-bonded pair of water molecules.

(a) the angle of the O–H$\cdots$O configuration has to be $\theta \geq 130°$, and

(b) the distance between two neighboring oxygen atoms has to be $d(\text{O–O}) \leq 0.330$ nm, and

(c) the hydrogen bond distance has to be $d(\text{O}\cdots\text{H}) \leq 0.240$ nm, and

(d) the interaction energy e_{HB} between the hydrogen-bonded water molecules should be more negative than -12.9 kJ mol^{-1}, and

(e) the occupancy of the σ^* orbitals has to be ≥ 0.0085.

Criterion (a) permits considerable bending of the hydrogen bond from linearity before it is regarded as broken. Criteria (a), (b), and (c) place geometrical constraints on the water molecules between which a hydrogen bond may exist. The values of $d(\text{O–O})$ and $d(\text{O}\cdots\text{H})$ are obtained from the pair correlation functions $g(\text{O–O}, r)$ and $g(\text{O–H}, r)$ defined in Section 3.1. Other cut-off distances have been used by some authors, however, some more lenient, some more restrictive. Criterion (d) is energetic, and some investigators allow a less negative binding energy, as low as -10 kJ·mol^{-1}. Considering the electron donor–acceptor nature of the hydrogen bond, the relevant molecular orbital of the acceptor water molecule is the σ^*_{OH} antibonding one that is empty in monomeric (nonhydrogen-bonded) water. The occupancy of this orbital, once hydrogen bonds are formed in liquid water, is the criterion to be used, but most investigators ignore criterion (e) for SCW.

Also, there is confusion regarding the term "the mean number of hydrogen bonds per water molecule." A water molecule may participate in up to four hydrogen bonds: two as a donor (of its hydrogen atoms) and two as an acceptor (to its oxygen lone pairs of electrons). However, there are necessarily two water molecules involved in each hydrogen bond. In the following (as throughout this book) the mean number of hydrogen bonds per water molecule in the system, symbolized by $\langle n_{HB} \rangle$, will be one half of the number of hydrogen bonds a water molecule has on the average, symbolized by χ (a symbol that has four dangling arms): $\langle n_{HB} \rangle = \chi/2$. The maximal value of $\langle n_{HB} \rangle$ is 2, that is, that of ordinary ice that has a regular tetrahedral hydrogen bond network ($\chi = 4$). Application of the listed criteria leads to $\langle n_{HB} \rangle \sim 1.7$ for water at ambient conditions [2], near enough to the value 1.73 for water at 25°C and 0.1 MPa, adopted by Hoffmann and Conradi [3] in their NMR study from the Raman results of Kohl et al. [4].

The fraction of water molecules in the system with $i = 0$, 1, 2, 3, and 4 hydrogen bonds is denoted by f_i. A quantity of interest concerning the extent of hydrogen bonding in SCW is the fraction f_0 of monomeric, that is, nonhydrogen-bonded, water. It is generally agreed that f_4 is negligible (≤ 0.02) in SCW at even the lowest temperature and highest density, but estimates of $f_0(T,\rho)$ (and other $f_i(T,\rho)$ values) vary widely between authors and estimation methods.

A further point to be considered here is the position (T,ρ) of the percolation threshold, beyond which no connected hydrogen-bonded chains exist anymore, but only small clusters (down to dimers) of water molecules. Again, the criteria for the existence of percolation need to be specified, as discussed in Section 3.1.2. A value $\chi = 1.55$ has been specified [5] for liquid water and employed in some studies of SCW, but other criteria have also been used.

3.1   DIFFRACTION STUDIES OF SCW STRUCTURE

X-ray and neutron diffraction measurements yield the intensities $I(\theta)$ of the beams diffracted from the sample at various angles θ for a fixed wavelength λ of the radiation. The angles are translated to the defined variable k

$$k = \lambda^{-1} 4\pi \sin(\theta/2) \tag{3.2}$$

to yield the functions $I(k)$. The structure factors $S(k)$ are then obtained as

$$S(k) \approx [I(k) - I(0)]/[I(\infty) - I(0)] \tag{3.3}$$

This expression uses the intensities measured at very small angles $I(0)$ and at very large angles $I(\infty)$ to normalize the intensities and eliminate the effect of the incoherent scattering on the relative intensities. Note that $\lim S(k \to 0) = \rho k_B T \kappa_T \sim 0.01$ (nominally $S(0) = 0$) where ρ is the number density of the diffracting atoms, and also that $S(\infty) = 1$. It should be mentioned that the structure factors obtained by both X-ray and neutron diffraction require corrections for extraneous effects, such as incoherent and nonelastic scattering, that may introduce uncertainties if not correctly applied.

The structure factor is related to the radial distribution function of the liquid $N(r)$ and the pair correlation function $g(r)$ is obtained from the experimental structure factors after application of a Fourier transform:

$$g(r) = (2\pi^2 \rho r)^{-1} \int_0^\infty (S(k)-1)k \sin(kr) dk \qquad (3.4)$$

Some uncertainty in the resulting $g(r)$ may be caused by truncation errors in the Fourier transform, and these must be guarded against.

The diffraction of X-rays takes place in the case of SCW only from the (electrons of the) oxygen atoms, so that it provides information concerning the structure as determined by these atoms, which correspond practically also to the centers of mass of the water molecules: $g(r) \approx g(\text{O–O}, r)$. Neutrons are diffracted from (the nuclei of) both oxygen and hydrogen atoms, but the hydrogen atoms may be exchanged, at least partly, for deuterium ones. In fact, the total pair correlation function on partial substitution of D atoms for H atoms in the water molecules is made up from contributions of three partial correlation functions:

$$g(r) = [b_O^2 g_{OO} + 4b_O b_D g_{OH(D)} + 4b_D^2 g_{HH(DD)}]/[(b_O + 2b_D)^2 \qquad (3.5)$$

where b_O and b_D are the scattering lengths of the O and D nuclei (prorated for the presence of H atoms) and the g_{OO} and so on are shorthand notations for the partial pair correlation functions: $g(\text{O–O}, r)$, $g(\text{O–H}, r)$, and $g(\text{H–H}, r)$. Hence, when the fractional substitution of H by D atoms is different in three distinct experiments at the same thermodynamic states, thereby changing b_D, neutron diffraction is capable of determining the partial pair correlation functions.

3.1.1 X-Ray Diffraction Studies of SCW Structure

Due to the limited information obtainable from X-ray diffraction, relatively few studies of SCW have been made by this method. Gorbaty and Dem'yanets [6–8] studied water up to 500°C and 100 MPa. They concluded that some hydrogen bonding persists in the supercritical range of temperatures. Yamanaka et al. [9] used an image plate X-ray diffractometer permitting rapid scans and applied it to water at various temperatures and just above the critical point: 649 K (376°C) and 80.4 MPa, at a density of 700 kg·m^{-3}.

A broad peak was seen in the $g(r)$ curve of the SCW, corresponding to the first coordination shell. It was analyzed in terms of a first peak at 0.292 nm with a coordination number $N_{co} = 1.6$, and another peak at 0.342 nm with a coordination number $N_{co} = 2.3$, altogether 3.9 molecules in the first neighbor shell. No further shells, that at ambient conditions have peaks at 0.45 and 0.67 nm, were observed above $\sim 140°C$. It was concluded that under supercritical conditions water clusters and small oligomers represent the spatial distribution of the water molecules. Okhulkov et al. [10] showed the function $r^2[g(r) - 1]$ of water at 100 MPa at temperatures from ambient up to 500°C in the region of the second peak, $r = 0.45$ nm, arising from tetrahedral hydrogen bonding and characteristic for ambient water. They concluded that the probability for tetrahedral configurations of the water molecules decreases up to 350°C but tends to grow as the temperature is increased. Gorbaty and Kalinichev [11] defined the quantity $f_4 = \chi/4$, the fraction of tetrahedral hydrogen bonding persisting in high temperature water ($f_4 = 1$ for ice). This can be calculated from the area under the 0.28 nm peak in the $g(r)$ function divided by the coordination number N_{co} that corresponds to the sum of the areas under the (resolved as Gaussians) 0.28 and 0.45 nm peaks. They found that f_4 decreased linearly from ambient to supercritical conditions:

$$f_4 = 0.851 - 8.68 \times 10^{-4} (T/K) \qquad (3.6)$$

up to 500°C and densities between 700 and 1100 kg·m^{-3} with an accuracy of ± 0.1 units. This interpretation of the X-ray diffraction data is consistent with infrared absorption spectral data (see Section 3.3).

Ohtaki et al. [12] reviewed the results obtained at the time from the X-ray diffraction from supercritical water, expanding on the above presentation of the situation, and compared them with some results from other methods of investigation. Recently Inui et al. [13] applied intense synchrotron radiation to study wide angle X-ray scattering from supercritical water at 40 MPa and 380–720°C (densities 610–100 kg·m^{-3}) and at 100 MPa and 385–827°C (densities 0.72–0.24 kg·m^{-3}). They reported only on the first peak and coordination numbers of the $g(r)$ curves, N_{co} increased linearly with the density from ~ 1.0 to ~ 3.5 in the experimental range, independent of the detailed pressure and temperature. They confirmed the earlier results of Gorbaty and Kalinichev [11] without adding new insight.

3.1.2 Neutron Diffraction Studies of SCW Structure

As mentioned above, much more detailed information is available from neutron diffraction measurements on SCW, provided that the neutron diffraction with isotope substitution (NDIS) method is employed. Postorino et al. first reported briefly the results of their study [14] and expanded on them subsequently [15]. They pointed out that in water at ambient conditions the partial pair correlation function $g(O-H, r)$ peaks at 0.19 nm and is indicative of the presence of strong hydrogen bonding. They obtained data in pure D_2O, equimolar $H_2O + D_2O$, as well as 0.003 $H_2O + 0.997$ D_2O at 300°C (9.5 MPa, 710 kg·m^{-3}) and 400°C (80 MPa, 660 kg·m^{-3}). They concluded that at 300°C this indicative peak is shifted to $\sim$0.21 nm and is reduced to about $1/3$ of its height at ambient conditions, but that at 400°C, that is, SCW, it disappeared altogether. Thus, they concluded, no hydrogen bonds between water molecules exist in SCW. Similar conclusions were drawn from the $g(H-H, r)$ peaks at 0.23 nm that first shifts (at 300°C) to 0.26 nm and becomes much less intense (at 400°C), indicating loss of orientational correlation between adjacent water molecules. Thus, in SCW no hydrogen bonding exists anymore. This conclusion was contrary to what was known at the time from other kinds of measurements, X-ray diffraction, spectroscopic and thermo-dynamic, as well as from computer simulation.

The controversy that arose finally settled on the accuracy of the neutron diffraction data and their interpretation. Bellissent-Funel et al. [16] then made some new measurements on D_2O at 380, 450, and 500°C with densities ranging from 230 to 730 kg·m^{-3} (note that the critical temperature for D_2O is lower than for H_2O, being 370.7°C). They did find the indicative 0.19 nm peak in the $g(r)$ in all their samples, and concluded that some degree of hydrogen bonding does persist in SCW (Fig. 3.2).

Furthermore, a neutron diffraction study by Soper et al. [17], a group that included some of the authors of the Postorino et al. paper [14, 15], employed a different neutron moderator and revised the earlier conclusions. Soper et al. used more adequate corrections for inelastic scattering to the experimental structure factors, and applied the new data to SCW at 400°C and 80 MPa (density 660 kg·m^{-3}). The 0.19 nm feature under these conditions is no longer a distinct peak but persists as a clear broad shoulder on the main $g(O-H, r)$ peak, but indicates the presence of some hydrogen bonding. If the geometric criterion for hydrogen bonding, of O–H$\cdots$O angles $\geq 150°$, is applied then still 42% of the hydrogen bonding in ambient water remains under the SCW conditions used. The main point, however, is that no extended network of hydrogen bonds exists in SCW [17], only dimers are to be considered to account for the hydrogen bonding in SCW.

New improved neutron diffraction measurements at 400°C and densities between 580 and 870 kg·m^{-3} were then made [18]. These still showed a shoulder on the lower side of the main $g(O-H, r)$ peak, but no feature at 0.45 nm, characteristic of tetrahedrally bonded water molecules (correspond-ing to a hydrogen-bonded network). It argues that the results from a single D/H

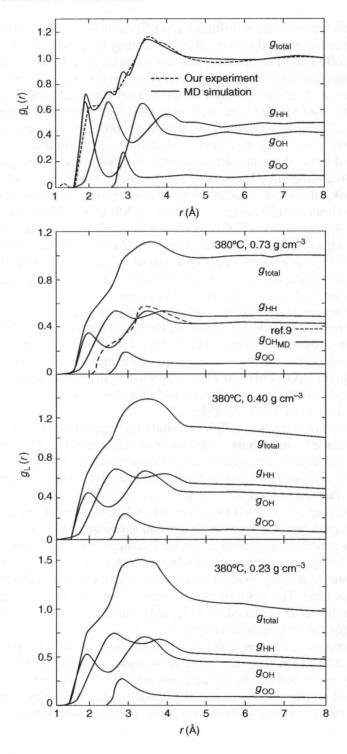

composition (i.e., pure D_2O) by Bellissent-Funel [16] could not provide partial pair correlation functions and the total $g(r)$ is not sensitive to the $g(O–H, r)$ hydrogen bonding signature peak [18].

To counter this criticism Bellissent-Funel et al. [19] carried out new neutron diffraction measurements on SCW consisting of D_2O, and mixtures substituted with 10% and 20% H_2O at 380°C with a density of 730 kg·m^{-3}. The H_2O content was limited in order to minimize errors due to nonequivalence of the incoherent and inelastic scattering corrections, but this reduced the variation of the weighting factor b_D for obtaining the partial structure factors from Eq. (3.5) and increased their uncertainties. The $g(O–H, r)$ resulting from these measurements did show a distinct small peak at 0.19 nm, rather than the shoulder in Ref. [18], but in general agreed with the latter results (obtained at 400°C). The new data also confirmed the absence of a feature at 0.45 nm in $g(O–O, r)$, showing the absence of tetrahedral hydrogen bonding. Bellissent-Funel summarized the findings from her group in Ref. [20] and compared them with results from computer simulations (Section 3.2) and Raman and NMR spectroscopy (Section 3.3).

The most recent contribution to this subject [21, 22] from the laboratories previously reporting on neutron diffraction by SCW [15, 17, 18] provided new measurements on SCW in three thermodynamic states: one gas-like (400°C, but at a density of 116 kg·m^{-3}) and two liquid-like (density 750 kg·m^{-3} and 400 and 480°C) and reconsidered also the [18] experimental results (at 400°C and a density of 580 kg·m^{-3}). The authors applied the isotope substitution method, using pure D_2O, pure H_2O, and equimolar mixtures to maximize the variation of the b_D coefficient in Eq. (3.5). They also applied the state-of-the-art treatment to the raw diffraction data and in addition applied the empirical potential structure refinement (EPSR) computer program for the analysis of the data. The resulting partial pair correlation functions of the liquid-like states confirm the earlier results, that is, the presence of the hydrogen bond signature peak at 0.19 nm in $g(O–H, r)$ and an absence of a feature at 0.45 nm in $g(O–O, r)$. However, closer examination of the EPSR data was interpreted in terms of the presence (at the liquid-like states) and absence (in the gas-like state) of percolation [23], that is, some hydrogen bonding network. The average number of hydrogen bonds of a water molecule in the gas-like state is only 0.8, below the percolation threshold of $\chi = 1.53$ [5],

---

FIGURE 3.2 The pair correlation functions in water (D_2O) at four states according to neutron diffraction data of Bellissent-Funel et al. [16] (reproduced by permission of the American Institute of Physics). (a) Liquid water at 25°C and ambient pressure, (b) SCW at just above the critical point ($T_r = 1.016$) and three densities as presented—the lowest panel at a density below the critical ($\rho_r = 0.714$).

that is exceeded in the liquid-like states, where it is 2 on the average. There occurs a maximum in the water molecule chain length distribution near 13 in these states, whereas a monotonic decrease occurs for the gas-like state. Some 5–10% of the molecules in the denser SCW do form four hydrogen bonds, so that it can be considered to be a percolating system.

Incoherent neutron scattering was used by Tassaing et al. [24] to obtain the diffusion coefficients in SCW at 380°C and six densities between 200 and 900 kg·m^{-3}, decreasing steadily as the density is increased. The hydrogen bond lifetime remained nearly constant near 0.19 ps up to a density of 600 kg·m^{-3} and decreased then to ~0.13 ps at 900 kg·m^{-3}, reflecting the higher incidence of collisions at the higher densities. Another recent report of measurements on very low density SCW (400°C, 6.8 kg·m^{-3}), employing neutron diffraction with isotope substitution came from another laboratory [25]. The results indicate that orientational correlation between neighboring water molecules is almost entirely lost and that small clusters involving a few water molecules are preferentially formed.

3.2 COMPUTER SIMULATIONS OF SCW

Computer simulations have been used as auxiliary data in experimental studies of SCW, for example, those using diffraction by X-rays or neutrons, spectroscopy, and dynamic processes, but many studies involving computer simulations standing on their own have also been made. The essential point to note is that the results depend on the potential function employed, functions that have evolved during the years and become more dependable. Also, the number of water molecules that are included in the simulations grew over the years, from 108 in 1980 [26] to 1000 in 2005 [23], yielding more representative results.

In the simulations, a specified number of water molecules are placed in a cubic box with periodic boundary conditions that simulate to some extent infinite systems, and in any case avoid problems with the surface of an ensemble of a limited number of molecules. Since the potential functions for water–water interactions involve Coulombic interactions between the partial charges on the atoms, the long range electrostatic effects are taken into account according to various schemes. Further details of these problems are presented in the papers quoted, for example, in Ref. [27].

The simulations are classified as Monte Carlo ones, in which the equilibrium structure of SCW is established, and molecular dynamics ones, in which in addition some dynamic features of SCW (self-diffusion, rotational orientations, vibrations, etc.) are also obtained. A point to be taken into account is that the potential function employed determines T_c, the virtual critical point of the system as derived from the simulations, which generally does not coincide with that of the *real* fluid, 374°C. It may be considerably lower, as low as 301°C (the TIP4T-FQ model) or 305°C (the SPC model), growing to 315°C (for the TIP4P model) and 336°C (for the BJH model), and approaching the experimental value, 350 or 367°C (for the SPC/E model), and 368 ± 6°C (for the TIP4T model). A critical temperature larger than the experimental one, $T_c = 437$°C, was found for a recently proposed model (Exp6-polar). This nonagreement is of particular importance for MD simulations, for which the temperature is a result of the calculations rather than an input condition as it is for MC simulations. Therefore, it is expedient to report $T_r = T/T_{c \text{ model}}$, the reduced temperature, rather than the nominal temperatures T reported by the authors.

3.2.1 Monte Carlo Simulations

In Monte Carlo simulations, a random configuration of a large number of water molecules at given temperature and pressure (*NPT* ensemble) or temperature and volume (*NVT* ensemble) is generated and the energy U of the system is computed from the intermolecular potentials appropriate for the water model employed. A molecule is then chosen at random and is moved randomly to a new position and orientation, and the energy of the system is again computed. If the energy change is $\Delta U \leq 0$ then the step is accepted directly; if $\Delta U > 0$ the Boltzmann factor $\exp(-\Delta U/k_B T)$ is compared with a random number between 0 and 1 then the move is accepted; if it is smaller and rejected otherwise [28]. A series of such steps establishes a Markov chain that is continued until equilibrium is reached, usually after 10^6 steps. The success of a Monte Carlo simulation depends critically on the adequacy of the intermolecular potential function employed for the calculation of ΔU. This function involves both the distance apart and the mutual orientations of any two of the N particles and the effects of third particles in their vicinity on the potential energy. The results of the simulation are the thermodynamic functions of the system as well as the pair correlation functions, as averages over the ensemble of configurations, assuming the ergodic principle of the equivalence of the time and ensemble averages.

The earliest application of MC simulations to SCW is that of O'Shea and Tremaine [26]. These authors applied the theoretical MCY-CI potential to an *NVT* ensemble of 108 water molecule and calculated the internal energy and isochoric heat capacity of SCW at three densities (374, 865, and 1000 kg·m^{-3}) at 600°C and at one density (1000 kg·m^{-3}) at 900°C. They refrained, however, from the calculation of the radial distribution or pair correlation functions, so did not obtain information on the structure of the SCW.

Further developments in the application of MC simulations to SCW are connected with Kalinichev, who published a series of papers on this subject [29–32]. Kalinichev started [29, 96] with a small *NPT* ensemble ($N = 64$, but showed similar results with $N = 216$ at the higher densities) and used the empirical TIP4P potential on SCW at 400°C and densities of 187, 390, 600, and 1050 kg·m^{-3} and at 500°C at densities of 257, 529, 1014, and 1262 kg·m^{-3}. Partial pair correlation functions: $g(O–O, r)$, $g(O–H, r)$, and $g(H–H, r)$ were shown, exhibiting the hydrogen bond signature peak at 0.19 nm in $g(O–H, r)$, the absence of the tetrahedral structural feature at 0.45 nm in $g(O–O, r)$, and hardly any rotational orientation correlation in $g(H–H, r)$. Kalinichev and Bass [30] used an NPT ensemble of $N = 216$ at 500°C and a wider range of pressures and densities (from gas-like at 10 MPa and 30.5 kg·m^{-3} to highly compressed liquid-like at 10 GPa and 1666.2 kg·m^{-3}) and the same TIP4P potential. The previous findings were confirmed and augmented with criteria for hydrogen bond formation: $d(O \cdots H) \leq 0.24$ nm for the hydrogen bond length and $e_{HB} \leq -10$ kJ·mol^{-1} for its energy (see Section 3.3). A combination of these criteria showed the average number of hydrogen bonds per water molecule $\langle n_{HB} \rangle$ increasing gradually from 0.1 at 10 MPa (30.5 kg·m^{-3}) to 1.25 at 10 GPa (1666 kg·m^{-3}). The average O–H$\cdots$O angle was 150° for the hydrogen bonds at all the thermodynamic states examined, being ~10° less linear than in ambient water. Further simulations were made [31] with an *NPT* ensemble of $N = 216$ and the TIP4P potential at 400°C and densities of 578, 660, 693, 731, 1050, and 1670 kg·m^{-3} as well as at 500 and 1000°C (at densities of 440 and 170 kg·m^{-3}, respectively). At 400°C the partial pair correlation functions at liquid-like densities differed considerably from the 1000°C functions both at liquid-like and gas-like densities. The hydrogen bond signature peak at 0.19 nm in the $g(O–H, r)$ at 400°C corresponded to $\chi = 2.1$ at 400°C, but it degenerated to a barely discernable shoulder at 1000°C yielding $\chi = 0.2$. Over the entire range from ambient to 1000°C the resulting average number of hydrogen bonds of a water molecule $\chi = f(T, P, \rho)$ can be shown to correlate well with the percentage of monomeric water molecules (those devoid of any hydrogen bonding):

$$100f_0 = 94.8 - 88.6\chi + 28.6\chi^2 - 3.0\chi^3 \qquad (3.7)$$

This quantity can be compared with values obtained from the equation of state (Section 2.1.3) and other sources (Section 3.4). The percolation threshold (see Section 3.3, $\chi = 1.55$) is exceeded only for temperatures below 600°C at any of the available densities.

Kalinichev and Churakov [32], using the same simulation conditions as above, dealt mainly with the structures of the water clusters in SCW that exist at conditions below the percolation threshold. Their maximal size appears to be 7 water molecules, with mainly chain-like trimers, tetramers, and penta-mers predominating over branched and ring structures. The distribution among such species and structures appeared to be substantially independent of the temperature and the pressure.

Matubayashi et al. [33] also employed the empirical TIP4P potential as well as the simple point charge (SPC) model for MC simulations with 648 water molecules, but used the dipole moment as a parameter in the range $1.85 \leq \mu \leq 2.35$ D. They studied SCW at 400 and 500°C at densities of 200, 400, and $600 \, \text{kg·m}^{-3}$, computing the expected NMR chemical shifts (see Section 3.3) and the partial pair correlation functions. The latter depend strongly on the dipole moments and $\mu \geq 2.15$ D is required to conform to results generally accepted for the pair correlation functions in SCW. The orientational structure in SCW is shown to be insensitive to the density, the appearance of a third water molecule near a hydrogen-bonded pair enhances slightly the probability of hydrogen bonding over a wide range of O–O–O angles ($>72°$) rather than peaking at the tetrahedral angle (109°) as in ambient water.

Jedlovszky et al. in a series of papers [23, 34, 35] studied the percolation properties of SCW by means of MC simulations. In the first of these [34], an *NVT* ensemble with $N = 250$ was used as were two potential functions: the rigid extended simple point charge (SPC/E) and a polarizable model (chang-ing from the *NVT* to an *NVE* ensemble). The SCW was at 400°C at a density of $652 \, \text{kg·m}^{-3}$. Small differences were only found between the results of the two potential functions and ensembles. The geometry of adjacent water molecules involved is shown in Fig. 3.1. The O–O–O angle in SCW may be as small as 36° compared with a minimum of 41° in ambient and hot water below the critical point. The O–H$\cdots$O angle in the SCW is $>135°$ for 67% of hydrogen-bonded pairs, and $>150°$ for 40% of the pairs, so that appreciable deviations from linearity occur. The average $\chi = 1.98$ in the SCW studied (above the percolation threshold), provided the cutoff O–H$\cdots$O angle is $>135°$, and the average cluster size is 249, with 0.53% of the water molecules monomeric, that differs considerably from the value predicted by Eq. (3.7), 8.2%. If the cutoff angle is increased to 150° then $\chi = 1.24$, below the percolation threshold. Thus, whether percolation (a hydrogen-bonded network) exists in SCW depends sharply on the geometric criterion employed to define a hydrogen bond.

Pátay and Jedlovszky [23] employed an *NPT* ensemble with 1000 water molecules and the SPC/E potential to study SCW. The thermodynamic states used were 430, 484, and 534°C at both gas-like densities ($17–28 \, \text{kg·m}^{-3}$) and liquid-like ones ($359–428 \, \text{kg·m}^{-3}$). Two water molecules were considered as hydrogen bonded when the distances $d(\text{O–O})$ and $d(\text{O}\cdots\text{H})$ are ≤ 0.35 and ≤ 0.25 nm, the respective minima in the $g(\text{O–O}, r)$ and $g(\text{O}\cdots\text{H}, r)$ curves, *or* when the hydrogen bond energy is $\leq -11.5 \, \text{kJ} \cdot \text{mol}^{-1}$. Two water molecules are considered to belong to the same cluster when connected by a chain of intact hydrogen bonds. The geometric criterion is more lenient, allowing percolating clusters at the liquid-like densities with a probability ≥ 0.5 (percolation threshold) even at the highest temperature studied, whereas the energetic one reduces the probability to nil even at 484°C. The average cluster size diminishes from 118 at 430°C to 35 at the higher temperature according to the geometric criterion. Pátay et al. [35] subsequently reexamined the 484°C isotherm by MC simulation with the *NVT* ensemble and SPC/E potential and 300 and 1000 water molecules. They used a series of closely spaced densities from 0.180 to 479 $\text{kg} \cdot \text{m}^{-3}$ in order to locate the percolation threshold. This was defined by the requirement that the fractal dimension of the largest cluster should exceed 2.53. According to this, only at densities $\geq 410 \pm 9 \, \text{kg} \cdot \text{m}^{-3}$ can the systems with both 300 and 1000 molecules be considered as percolating. However, $\chi = 1.55$ previously considered as the threshold of percolation [5], is exceeded already at a density of 344 $\text{kg} \cdot \text{m}^{-3}$. This point has not been commented on.

3.2.2   Molecular Dynamics Simulations

In molecular dynamics computer simulations the trajectories of the particles of the system (usually the *NVE* ensemble, at constant volume and total energy of the system) are followed, starting from random positions, orientations, and linear and angular velocities in the simulation box, according to the classical equations of motion over a short time step (0.1–1 fs). The positions, velocities, v, orientations, and angular velocities of each particle, resulting from the application of pairwise interaction potentials, are integrated over this short period, and the history of the system is collected over 10^3 to 10^5 steps. The temperature of the system is given by the average kinetic energy of the N particles of mass m [28]:

$$T = (2/3Nk_\text{B}) \left\langle \sum\nolimits_N (mv^2/2) \right\rangle \tag{3.8}$$

The water models and potentials used for MD simulations yielded various values for the critical temperature of water, T_c (see above).

Mountain [36] and Cummings et al. [37] were among the first who applied MD simulations to supercritical water. Mountain [36] used the TIP4P water model potential on 108 water molecules and derived the partial pair correlation functions $g(\text{O–O}, r)$, $g(\text{O–H}, r)$, and $g(\text{H–H}, r)$ for several SCW states. He adopted the geometrical criterion for the existence of a hydrogen bond: $d(\text{O} \cdots \text{H}) \leq 0.24$ nm and reported the average number of hydrogen bonds per water molecule, $\langle n_{HB} \rangle$. At the SCW states $\langle n_{HB} \rangle$ decreases mildly with increasing temperatures, up to $T_r = 2.34$ but strongly with diminishing densities. At $T_r = 1.36$ $\langle n_{HB} \rangle$ is 1.36 at 999 kg·m^{-3}, 0.83 at 600 kg·m^{-3}, but only 0.40 at 250 kg·m^{-3}. Cummings et al. [37] used the SPC model potential on 216 water molecules at $T_r = 1.0$ (the critical temperature for this model) and a density of 405 kg·m^{-3} and on 255 molecules at $T_r = 1.05$ and 270 kg·m^{-3} ($\rho_r = 1.0$ for this model). Agreement of the calculated bulk properties with the experimental ones of water was only fair, but the resulting partial pair correlation functions did resemble those from neutron diffraction discussed above (Section 3.1.2). The rigid point charge Coulombic and Lennard-Jones pair potential of the SPC model, with empirically adjusted parameters for ambient water, is seen to be inadequate for describing dependably the properties of SCW.

Kalinichev [38] employed instead the flexible Bopp–Janscó–Heinzinger (BJH) model of water (although the pressures calculated by the BJH potential far exceeded the experimental ones), applying it to an NVE ensemble of 200 water molecules at four SCW states: one gas-like ($T_r = 1.19$ and 166.6 kg·m^{-3}) and three liquid-like ($T_r = 1.49$ at 528.2 and 1284.0 kg·m^{-3} and $T_r = 1.21$ at 971.8 kg·m^{-3}). The simulations yielded self-diffusion coefficients in fair agreement with experimental values, but no structural information was provided. In a later paper, Kalinichev and Heinzinger [39] applied the same BJH potential to the above systems to deduce the vibrational spectra, but again, no structural information was presented. Kalinichev and Churakov [40, 41] again employed the BJH flexible potential to the above states of water and reexamined the topological properties of the clusters. Large linear clusters, of up to 10 molecules, exist even at the lowest densities studied. In high density SCW the percolation threshold is exceeded and infinite clusters are formed. Krishtal et al. [42] summarized the finding from the use of the BJH potential function for MD simulations of SCW at $1.00 \leq T_r \leq 1.21$ and densities between 200 and 600 kg·m^{-3}. They employed both geometric and energetic criteria for the hydrogen bonding and presented values for $\langle n_{HB} \rangle$ diminishing from ~1.3 to ~0.6 as the temperature increased (read from a small scale figure) according to the geometric criterion. Also, the fractions of water molecules with 0, 1, 2, 3, and 4 hydrogen bonds were presented in a small scale figure, from which it is seen that practically no molecules with 4 bonds exist but a small fraction (10% down to 2%) with 3 bonds exists. The general conclusion was that on passing the critical point the hydrogen bond network collapses and hydrogen-bonded clusters of various sizes appear.

Mizan et al. [27] discussed critically the choice of the water model potential function for MD simulations of SCW. They showed that rigid models, such as SPC or TIP4P, yield critical points far from the experimental one, whereas flexible models succeed much better in reproducing the thermodynamic properties. But among various flexible models there are still variations and the Teleman–Jönsson–Engström (TJE) model appeared to yield the best potential for simulating SCW (but T_c was not specified). The authors applied this potential to 256 water molecules at nominally 500°C at four densities: 115.3 (gas-like), 257.0, 405.8, and 659.3 kg·m^{-3}. The partial pair correlation functions for these states were compared with those of water at ambient conditions. The $g(O–O, r)$ of the gas-like SCW showed no second peak and those of the denser fluids showed only a hint of it at $r > 0.5$ nm. The 0.18 peak was, however, clearly shown by the $g(O–H, r)$ curves of all the SCW states, but its intensity appeared to diminish with increasing density. Self-diffusion coefficients calculated from the simulations agreed well with experimental values. Dang [43] also evaluated the importance of polarization effects in the potential functions employed, comparing for SCW at $T_r = 1.09$ and 660 kg·m^{-3} the partial pair correlation functions obtained with the rigid SPC/E model and a rigid four-site polarizable model. The latter showed better agreement with the neutron diffraction-with-isotope-replacement experimental results, except that it confirmed the presence of a significant hydrogen bonding signature peak in $g(O–H, r)$ at ~0.2 nm.

Liew et al. [44] used a flexible four-site water model, similar to the rigid TIP4P one but with modified parameters, for MD simulations on an *NVT* ensemble of 500 water molecules. This model was capable of near reproduction of the critical temperature ($T_c = 368 \pm 6$°C) and density (307 ± 5 kg·m^{-3}) of real water, contrary to rigid models. It also reproduced well the experimental diffusion coefficient at $T_r = 1.09$. The study dealt in particular with the three-dimensional structure, a feature that cannot be seen in the orientation-integrated $g(O–O, r)$ curves. Bifurcated hydrogen bonds between adjacent water molecules were found beside the linear ones. It was argued that the kinetic energy of the molecules at 400°C, ~8.4 kJ·mol^{-1}, is sufficient to form bifurcated hydrogen bonds that are less stable than the linear ones by only ~5.8 kJ·mol^{-1}.

Yoshii et al. [45] used a polarizable SPC model, that yielded a critical temperature ($T_c = 288$°C) much below the experimental one, so what they reported for nominally 327°C corresponds to a reduced temperature $T_r = 1.07$, that is, to SCW at 400°C. They performed MD calculations on 256 water molecules and used $d(O \cdots H) \leq 0.25$ nm as a definition of an intact hydrogen bond. They reported the partial pair correlation functions for six densities from 27 to 1000 kg·m^{-3}. The coordination numbers were obtained from integration of $g(O–O, r)$ curves up to 0.45 nm and increase steadily with the density, up to ~12 at the highest density. The $\langle n_{HB} \rangle$ values also increase steadily with the density, reaching 1.9 at the highest density, the hydrogen bonds being bent at 21° from linear on average. The χ is larger than the percolation threshold of 1.55 at densities above 800 kg·m^{-3}. Petrenko et al. [46] proposed the use of yet another potential function, namely the TIP4P modified for the explicit formation of hydrogen bonds. MD simulations with 216 molecules at various temperatures along the 50 MPa isobar as well as at $T_r = 1.0$ and 80 MPa and at $T_r = 1.27$ and 100 MPa. The results were essentially the same as for the simulations by others. Along the 50 MPa isobar the average number of hydrogen bonds of a water molecule, χ decreased from ~2.0 at $T_r = 1.0$ to 0.45 at $T_r = 1.27$. Still another potential function, Exp6-polar, was suggested by Bastea and Fried [47] that was similar to the SPC one but better applicable to high-temperature SCW. Simulations with this potential reproduced well the experimental diffusion coefficients at $T_r = 1.37$ and the relative permittivities at $T_r = 1.09$.

Dynamic aspects of SCW have also been simulated by MD. Yoshii et al. [45] with their polarizable SPC model found the rotational relaxation time to be minimal, $\tau_{rot} \sim 0.04$ ps, at a density of 147 kg·m^{-3} and it increases to $\tau_{rot} \sim 0.10$ ps at 1000 kg·m^{-3}. Skaf and Laria [48] studied the dielectric relaxation by means of MD simulations with SPC/E water at a variety of SCW states. The Debye relaxation times, τ_D, increased monotonically with increasing densities (and diminishing temperatures) in the range 0.21–0.44 ps, but fail to reproduce the experimental increase in τ_D at densities below 400 kg·m^{-3}. Matubayashi et al. [49] studied the rotational dynamics of SCW at $T_r = 1.33$ with the flexible water model TIP4P-FQ (having a critical temperature as low as 301°C!) with 256 molecules. The reorientational correlation times, τ_{2R}, increased from 0.048 ps at 100 kg·m^{-3} to 0.064 ps at 600 kg·m^{-3}, but depended on the assumed quadrupole coupling constant (varying between 256 and 308 kHz). The values of τ_{2R}, representing the motions of single molecules, are expected to be lower than those of the dielectric relaxations times, τ_D, depending on the collective motion of several molecules.

Guardia and Marti [50, 51] used the SPC/E water model for 256 molecules for their MD simulations of SCW reorientational dynamics at $T_r = 1.016$. They studied a wide range of densities at this temperature, from 10 to 700 kg·m^{-3}, and examined the orientational correlation times along four vectors. Along the dipole vector τ_1 increased steadily from 0.027 to 0.225 ps, whereas τ_2 showed a minimum of 0.031 ps at 50 kg·m^{-3}. Along the O–H bond and the line connecting the two hydrogen atoms the times also showed minima at this density, and in a direction perpendicular to the plane of the molecule the minimum occurred at 100 kg·m^{-3}. At the highest density, 700 kg·m^{-3}, the times were ~0.20 ps for τ_1 and 0.08 ps for τ_2 at all four directions. The dielectric relaxation times τ_D were about twice longer, except at the lowest densities. In their later paper [51], they added data at $T_r = 1.125$ and $T_r = 1.36$, showing the expected faster reorientation times as the temperature was increased, but essentially the same qualitative results as at $T_r = 1.016$.

Dyer and Cummings [52] applied MD simulation to two water models: the Gaussian charge polarizable model and the Car–Parrinello *ab initio* model to 256 and 32 water molecules, respectively at two SCW densities: 1000 and 600 kg·m^{-3} over the temperature range 423–723°C. They calculated the total dipole moment and the radial distribution function and estimated the number of hydrogen bonds per water molecule, χ. The values of $\langle n_{hb} \rangle = \chi/2$ interpolated from a figure are shown in Table 3.1.

The local density inhomogeneities and their dynamics were studied in SCW at $T_r = 1.03$ over a wide density range: 61–644 kg·m^{-3} by MD simulations with the *NVT* ensemble of 500 molecules and the SPC/E model by Skarmoutsos and Samios [56]. First shell density enhancement of 43% and second shell enhancement of 21% were found at a density of 122 kg·m^{-3}, the relative quantities diminishing at increasing densities. The density rearrangement times after a perturbation, $\tau_{\Delta\rho}$, decreased from $\sim$1.3 ps at the lowest to $\sim$0.3 ps at the highest density examined. Skarmoutsos and Guardia [57] again used this model for MD simulations of 500 water molecules at the above states to calculate the lifetimes of hydrogen bonds, defined geometrically: d(O–O) ≤ 0.36 nm and d(O$\cdots$H) ≤ 0.24 nm. The continuous lifetime, τ_{HBc}, remaining near 0.075 ps over the density range and the intermittent lifetimes, τ_{HBi}, decrease from 0.62 ps at 61 kg·m^{-3}, to a minimum of 0.44 ps near the critical density and increases to 0.47 ps at 644 kg·m^{-3}. The reorientation correlation times depend on the number of hydrogen bonds a water molecule is engaged in and the density, increasing with them, as expected.

Local densities were also stressed by Kandratsenka et al. [58] as being more relevant to the hydrogen bond connectivity than the bulk density. MD simulations were made with the SPC/E model potential on an *NVT* ensemble of 108 water molecules at $T_r = 1.09$ at 428 kg·m^{-3} and the g(O–O, r) curve was resolved into contributions from first, second, third, and fourth neighbors. The d(O–O) distances were then correlated with the maxima of the O–H stretching vibrations of dilute HOD in D_2O. Swiata-Wojcik and Szala-Bilnik [54] also used MD simulations with the BJH flexible model to study the inhomogeneity in SCW (at $T_r = 1.10$ and 1.04) with results shown in Table 3.1 for $\langle n_{HB} \rangle$ (larger than other estimates) and in Table 3.3 for f_0 (in agreement with previous estimates).

3.3 SPECTROSCOPIC STUDIES OF SCW

3.3.1 Infrared Absorption Spectroscopy

Infrared spectroscopy applied to water reveals mainly the asymmetric O–H stretching vibration ν_3 and secondarily also overtone and combination bands

TABLE 3.1 The Average Number of Hydrogen Bonds per Water Molecule, $\langle n_{HB} \rangle$, Interpolated from Data at Reported Temperatures and Densities

ρ/kg·m^{-3}	t/°C				
	380	400	450	500	600
100		0.09[a]	0.08[a]	0.07[a]	0.05[a]
		0.20[b]	0.15[c]	0.14[d]	0.22[e]
		0.32[e]		0.22[e]	
150		0.18[a]	0.14[a]	0.11[a]	0.09[a]
		0.35[b]		0.25[f]	
		0.35[f]			
		0.44[g]			
250		0.27[a]	0.25[a]	0.22[a]	0.35[f]
		0.52[f]	0.30[c]	0.29[d]	0.39[e]
		0.56[b]		0.39[f]	
		0.64[e]		0.48[e]	
350	0.45[h]	0.40[a]	0.36[a]		
		0.67[b]			
		0.81[g]			
450	0.52[h]	0.47[a]	0.49[a]	0.47[d]	
	1.06[g]	0.71[f]		0.61[i]	
		0.72[b]		0.24[i]	
600		0.72[b]	0.26[i]	0.81[f]	
		0.88[f]			
		0.32[i]			
		0.95[j]			
		0.60[k]		0.53[k]	0.48[k]
		1.26[g]			
750		0.52[l]	0.44[l]	0.36[l]	0.97[e]
		0.75[b]		0.71[d]	0.97[f]
		1.08[f]		1.01[f]	
		1.24[e]		1.07[e]	
		1.10[j]		1.10[j]	
1000		1.00[k]		0.93[k]	0.87[k]

[a] From NMR chemical shifts [3].
[b] From IR absorption [53].
[c] From MD simulation [45], at 430 rather than 450°C.
[d] From MC simulation [30].
[e] From EoS calculations [94G].
[f] From MD simulation [36].
[g] From MD simulation [54].
[h] From IR absorption [55] but at densities of 325 and 425 kg·m^{-3}.
[i] From X-ray diffraction [1] at a densities ρ/kg·m^{-3} of 688 (400°C), 612 (450°C), and 512 (500°C).
[j] From neutron diffraction [21] at 580 kg·m^{-3} at 400°C and at 480 rather than 500°C at 750 kg·m^{-3}.
[k] From MD simulation [52].
[l] From X-ray diffraction [11] at densities 700–1100 kg·m^{-3}.

involving the symmetric O–H stretching mode v_1 and the H–O–H bending mode v_2, as well as rotatory structure of the vibrational bands. All these are sensitive to the surroundings of the excited water molecule, including its hydrogen bonding with its neighbors.

Luck [59, 60] was among the first who applied infrared absorption spectroscopy to near-critical and supercritical water. The measurements extended up to 390°C, and it was found that the $8749 \, cm^{-1}$ $2v_3 + v_2$ combination band, characteristic of vibrations of free O–H groups (those in molecules that do not participate in any hydrogen bonding) has a small intensity maximum at the critical point. This $2v_3 + v_2$ combination band of a given free O–H of a monomeric water molecule is somewhat more intense beyond 350°C than the $8696 \, cm^{-1}$ band of the O–H vibration of a molecule in which the other OH group does participate in a hydrogen bond. Luck and Ditter subsequently [61] extended these studies to dilute (5.2 mol%) HOD in D_2O and presented the IR spectrum at 386°C, assigning the overtone $2v_3$ at $7163 \, cm^{-1}$ to the free O–H vibration. Small shifts of the 6250 and $7200 \, cm^{-1}$ bands to higher frequencies were observed in SCW up to ~392°C. They calculated the fraction of free O–H groups from various relevant IR bands from low temperatures up to just beyond the critical point, arriving at 100% such groups, that is, $f_0 = 1$.

Franck and Roth [53] used 8.5 mol% HOD in H_2O (rather than in D_2O) and measured the IR spectra at 400°C at various pressures (densities of SCW from 150 to $900 \, kg \cdot m^{-3}$). They noted a gradual shift of the $2600 \, cm^{-1}$ band of the O–D vibration to lower frequencies but larger integrated intensities as the density was increased. The authors concluded that almost all the O–D groups are hydrogen bonded to some extent at densities $\geq 100 \, kg \cdot m^{-3}$, contrary to the findings of Luck [60].

Kalinichev and Heinzinger [39] quoted experimental results of Gorbaty [62] for the v_1 O–H stretching frequencies of water in three states ($t/°C$, $\rho/kg \cdot m^{-3}$): (357, 693), (400, 167), and (499, 528) (but note that the temperatures are nominal, obtained by molecular dynamics simulation with the BJH potential function that yielded $T_c = 609 \, K$ instead of 647 K). The frequencies were 3569, 3600, and $3580 \, cm^{-1}$, respectively, agreeing within $70 \, cm^{-1}$ with the frequencies calculated by the MD simulations. The latter also provided values for the v_2 H–O–H bending and v_1 and v_3 stretching bands, as also for two further SCW states: (407, 972) and (498, 1284). The frequencies of the bending mode band near $1600 \, cm^{-1}$ increase with the density but those of the stretching modes near $3600 \, cm^{-1}$ decrease with increasing densities.

More recently Bondarenko et al. [63] measured the IR spectra of 3 mol% HOD in H_2O at the 100 MPa isobars at 400, 450, and 500°C, showing a blue shift with decreasing intensities of the 2600 band for O–D stretching at increasing temperatures. Kandratsenka et al. [08B] measured IR spectra of 2 mol% HOD in D_2O at one SCW state: 397°C and 252 kg·m^{-3}. They confirmed the findings of Franck and Roth [53] and related the MD-calculated band 3600 band frequencies to the *local* d(O–O) distances and in turn to the number of hydrogen bonds $\langle n_{HB} \rangle$. As the former increase and the latter decrease the band maximum shifts to higher frequencies. Tassaing et al. [55] measured the IR spectrum of water just above the critical point, that is, at 380°C, confirming the features reported by Gorbaty and Bondarenko [64] measured at 500°C. The resulting spectra were deconvoluted into bands corresponding to small clusters: monomer ($\equiv$ free O–H), dimer, and linear trimer and the fractions of each species were plotted as a function of the density, reaching f_0 ~63% (down from 100%), f_1 ~21%, and f_2 ~16% as the density increased from 0 to 430 kg·m^{-3}.

3.3.2 Raman Scattering Spectroscopy

Raman spectroscopy appears to have been more widely applied to SCW and to produce more meaningful results than IR measurements. When applied to water the Raman spectrum reveals the symmetric O–H stretching vibration v_1 and is much less sensitive to the asymmetric stretching vibration v_3 and the H–O–H bending mode v_2.

Kohl et al. [4] seem to be the first to apply Raman spectroscopy to SCW containing 9.7 mol% HOD in H_2O; they did so for 400°C and a variety of densities between 40 and 800 kg·m^{-3}. The 2730 cm^{-1} O–D v_1 stretching vibration band was deconvoluted into two components, corresponding to hydrogen-bonded and nonhydrogen-bonded water molecules, indicating that only 33% of the water molecules are hydrogen bonded. Frantz et al. [65] presented Raman spectra at near the critical point, 376°C, and three SCW temperatures: 401, 451, and 505°C at various pressures, from 23 to 202 MPa. The sharp peak of the 3600 cm^{-1} band at the lowest pressure broadens and shifts to lower frequencies as the pressure is increased along the isotherms. The favored interpretation of the results was in terms of increasing fractions of non- or weakly hydrogen-bonded water molecules with frequencies $\geq$3550 cm^{-1} as the temperature is increased and/or the density is decreased.

Walrafen has dealt extensively with the Raman scattering from water over a range of temperatures and pressures and has also dealt with SCW. The 3250 cm^{-1} shoulder of the Raman band was assigned to correlated collective motions of hydrogen-bonded molecules and the 3400 cm^{-1} peak to uncorrelated stretching vibrations of the O–H bonds. Walrafen and Chu [66] discussed the relation between the intensity of the correlated Raman scattering and the structural correlation lengths (*SCL*) obtained from diffraction measurements, and involved one SCW state, that at 400°C but an unspecified density $\geq 800 \, \text{kg} \cdot \text{m}^{-3}$. Over a wide temperature range, including the SCW state, the intensity ratio, $R_{c/u}$, of the correlated to the uncorrelated bands is linear with the structural correlation lengths: $R_{c/u} = -0.304 + 2.55 \, (SCL/ \text{nm})$. The linearity persists to SCW suggesting that hydrogen-bonded clusters are present there, which have some degree of collective motions of their molecules.

Masten et al. [67] extended the Raman spectral measurements on water to 430°C but showed the v_1 O–H stretching frequencies of the 3600 cm^{-1} peak for just above the critical point, that is, at 380°C. Ikushima et al. [68] extended the Raman spectral measurements on SCW to 510°C and quoted a value from Matsen et al. [67] for 400°C, not appearing in the original publication. Figure 3.3 drawn following Walrafen et al. [69] illustrates the frequency shifts near the critical point. As the temperature was raised a blue shift Δv_1 was seen, indicating the breaking of hydrogen bonds, but v_1 did not reach the monomer value of 3657 cm^{-1}, so that some hydrogen bonding persisted. The cooperative network of hydrogen bonds in SCW at densities $> 400 \, \text{kg} \cdot \text{m}^{-3}$ is considered to be absent and only dimers are predominant species for the hydrogen bonding. The results are compared with the NMR results of Hoffman and Conradi [3] (see below).

Carey and Korenowski [70] deconvoluted the v_1 band for SCW at 400°C and 25.6 MPa into five Gaussian components: three at 3314, 3486, and 3522 cm^{-1} corresponding to hydrogen-bonded O–H and two at 3586 and 3658 cm^{-1} to

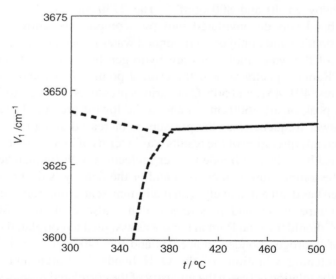

FIGURE 3.3 Raman frequencies of the symmetric O–H stretching vibrations, v_1, in water near the critical point, after Ref. [69]. Short dashed curve: saturated water vapor, long dashed curve: liquid water, and continuous curve: SCW.

nonhydrogen-bonded O–H. From the ratio of the sums of the integrated intensities of these two kinds of bands the enthalpy of hydrogen bonding was determined (from data at $\leq 200°C$) as $-10.6 \pm 0.4\,kJ \cdot mol^{-1}$. According to Yasaka et al. [71] the Raman frequencies of SCW at 400°C were red-shifted linearly as the density increased from 200 to $600\,kg \cdot m^{-1}$, as a result of hydrogen bonding. The same was noted by Yui et al. [72], in whose study of water at 400°C the pressure was increased from 19.5 to 36.7 MPa. Except near the critical point, the frequency of the O–H stretching vibration is $\nu_1/cm^{-1} = 3645.5 - 0.03831\,(\rho/kg \cdot m^{-3})$. Near the critical point density fluctuations affect the frequency, causing blue-shifts $\Delta\nu_1$ up to $25\,cm^{-1}$. A very short and intense laser pulse [73] at the ν_1 band frequency has the same effect as heating water to supercritical conditions: hydrogen bonds of the excited water molecule are disrupted, to be reformed with an appropriate relaxation time.

Rather than using light Raman scattering spectroscopy, Wernet et al. [74] used X-ray Raman scattering at the oxygen K-edge and applied it to SCW at 380°C and a density of $540\,kg \cdot m^{-3}$. The resulting spectrum showed a near-edge peak at 534 eV, a shoulder at 536 eV, and a major contribution between 537 and 542 eV (post-edge). The former two features correspond to those found in dilute water vapor, hence are characteristic of nonhydrogen-bonded water molecules. The intensity of the post-edge region, an energy region absent in water vapor, points to a large degree of hydrogen-bonding. A comparison with the corresponding spectrum of ice shows a shift of 3.3 ± 0.1 eV toward lower energies of the second post-edge feature, corresponding to $d(O–O) = 0.32$ nm. Clusters of 5–10 water molecules represent the hydrogen-bonded domains in the just-critical water studied. The fraction of water molecules with two nonhydrogen-bonded O–H groups, f_0, was estimated from the pre-edge features as $35 \pm 20\%$. This was construed to mean that $65 \pm 20\%$ of the water molecules have four hydrogen bonds, although elongated and bent, with $d(O \cdots H)$ between 0.165 and 0.230 nm and angles O–H$\cdots$O between 150° and 180°.

3.3.3 Nuclear Magnetic Resonance

Jonas et al. [75] claimed to be the first who applied (proton) NMR to SCW. They reported the spin-lattice relaxation times T_1 at 400–600°C and fairly low densities, $50–350\,kg \cdot m^{-3}$. The T_1 increased with increasing densities but diminished with increasing temperatures. They were unable to obtain the angular momentum correlation times τ_J from their data but reported Enskog relaxation times (valid for nonpolar gases as hard spheres), from 0.2 to 2.8 ps, which were some five times longer than estimated τ_J values. The shortness of the latter was ascribed, in part, to strongly anisotropic forces between water molecules (i.e., hydrogen bonding but not specified as such).

A more detailed study by Lamb and Jonas [76] followed, in which the spin lattice relaxation times T_1 were measured up to 700°C, but with emphasis on the region near T_c. The data for the lowest SCW temperature, 400°C, extended to 750 kg·m^{-3} but the maximal densities studied diminished at 500 and 600°C and were only 350 kg·m^{-3} at 700°C. The estimated angular momentum correlation times τ_J were now reported, increasing with the increasing temperatures at a constant density (at 350 kg·m^{-3} from 0.0641 ps at 400°C to 0.103 ps at 700°C) but decreasing with the density at a constant temperature (from 0.777 ps at 500°C and 50 kg·m^{-3} to 0.0468 ps at 600°C and 350 kg·m^{-3}). The reorientation correlation times τ_θ were assumed to be somewhat lower than the angular momentum correlation time τ_J, and were estimated to be in the range 0.008–0.060 ps.

Hoffmann and Conradi [3] shifted the attention from relaxation times to the chemical shifts measured by proton NMR in SCW at 400–600°C, using dilute benzene as an internal reference. The water resonance shifts to lower frequencies (i.e., more shielded) at increasing temperatures where the hydrogen bond network is increasingly destroyed. As the density approaches zero, the chemical shift converges to $\delta = -6.6$ ppm at all temperatures, a value taken to represent monomeric (nonhydrogen-bonded) water. The results, compared with the chemical shift $\delta = -2.5$ ppm for water at 25°C and 0.1 MPa yielded the linear relationship of the relative extent of hydrogen bonding η (relative to ambient water) as

$$\eta = 1.610 + 0.2439\delta \tag{3.9}$$

The experimental chemical shift is the weighted average between the values for hydrogen-bonded and nonhydrogen-bonded water molecules. Applied to SCW at 400°C and 520 kg·m^{-3}, for instance, the chemical shift of $\delta = -5.4$ ppm corresponds to 29% of hydrogen bonding relative to ambient water.

Matubayashi et al. [77, 78] measure the proton chemical shifts in SCW at 380, 390, and 400°C at the densities, 290, 410, 490, and 600 kg·m^{-3}. Magnetic susceptibility corrections were applied in order to refer the chemical shifts to an isolated water molecule. Over this narrow temperature range δ is only weakly temperature dependent, but increases with the density. The resulting δ values were in good agreement with those of Hoffmann and Conradi [3], and both sets of authors stressed that no cooperativity of hydrogen bonding needs to be taken into account for SCW. In a subsequent paper Matubayashi et al. [49] reverted to a study of the relaxation times, however, for supercritical D_2O. In this case the relaxation is governed by the quadrupolar relaxation mechanism, providing a second order reorientational time τ_{2R}. They measured the spin-lattice relaxation times T_1 for D_2O at 400°C at densities of 0.1–0.6

relative to those of the ambient liquid. The derived values of τ_{2R} ranged from 0.047 to 0.082 ps, increasing with the density.

More recently Yoshida et al. [79, 80] used proton NMR with a high resolution (500 MHz) instrument to study SCW at 400°C. The relaxation time of the translational velocity, τ_D, is related to the self-diffusion coefficient D: $\tau_D = MD/RT$, where M is the molar mass (but this is a dimensionally wrong equation!). The solvation relaxation time, τ_S, is the relaxation time in the solvation shell of a given water molecule. The latter declines mildly with the number n of molecules in this shell, from 0.04 ps at $n = 4$ to 0.01 ps at $n = 22$ at the supercritical density of 600 kg·m^{-3}, corresponding to n_{HB} of molecules from 0 to 8 (not per molecule but in the solvation shell or cluster). The ratios τ_D/τ_S first increase up to densities of 400 kg·m^{-3} and then fall again up to 1000 kg·m^{-3}, but are > 1 at all the densities, indicating more gas-like than liquid-like behavior of the relaxation mechanism.

Yamaguchi et al. [81] calculated the proton chemical shift theoretically from several *ab initio* molecular orbital theories and concluded that at the critical temperature and density water is made up from 80% monomer and 20% dimer. Similar calculations were reported by Sebastiani and Parrinello [82] for just critical water (374°C, 320 kg·m^{-3}) and SCW slightly above the critical point (380°C, 730 kg·m^{-3}). There are 14% and 37% hydrogen-bonded water molecules relative to ambient water under these conditions.

All these studies pertain to proton NMR (or ^{2}H-NMR [49]) and only Tsukahara et al. [83] employed ^{17}O NMR in order to study SCW, up to 425°C, reporting both chemical shift, δ, and spin lattice relaxation times, T_1. Ambient water was used as an external reference and magnetic susceptibility corrections were applied. Low density water vapor has $\delta = -36.1$ ppm against the reference, so that the extent of hydrogen bonding relative to ambient water was taken, in analogy with Eq. (3.9) of Hoffmann and Conradi as

$$\eta = 1 + 0.0274\delta \qquad (3.10)$$

Thus, $\eta = 0.40$ at T_c and 480 kg·m^{-3}, falling to 0.17 at 430°C and 240 kg·m^{-3} (read from a figure). The T_1 values depend more on the temperature than on the density and increase from 0.16 s at T_c to 0.26 s at 430°C. These times are dominated by the quadrupole relaxation that has an activation energy of 15.6 kJ·mol^{-1}.

In conclusion, it should be stated that most authors applied NMR to SCW at or just above the critical point (i.e., 400°C). Only Hoffmann and Conradi [3] extended the temperature to 600°C, and the relationship between the proton chemical shift and the average number of hydrogen bonds of a water molecule existing in SCW, Eq. (3.9), has been of great use to other researchers.

3.3.4 Dielectric Relaxation Spectroscopy

The dynamics of the water molecules in SCW has been studied by dielectric relaxation spectroscopy among other methods, such as NMR and molecular dynamics simulations. Some aspects of this have already been presented in Section 2.3.2. The Debye theory expression relates the permittivity ε of a fluid in an alternating field to its frequency ω (cf. Eq. (2.24)) in terms of the Debye relaxation time τ_D. Okada et al. [84] presented detailed tabulated experimental τ_D data up to $\omega = 40\,GHz$ for both supercritical H_2O and D_2O up to $\sim 600°C$ and densities $\leq 700\,kg \cdot m^{-3}$, that are reproduced in part in Table 3.2. The values for D_2O are some 30% larger than those for H_2O at a given temperature and density.

These authors and Yao and Hiejima [85] introduced the hydrogen bonding of a fraction f_b of the water molecules in SCW into the relaxation expression as follows:

$$\varepsilon(\omega) = \varepsilon(\infty) + (\varepsilon(0) - \varepsilon(\infty))[(1 - f_b)/(1 + i\omega\tau_{D\,f}) + f_b/(1 + i\omega\tau_{D\,b})]$$

(3.11)

The measured relaxation time is $\tau_D = (1 - f_b)\tau_{D\,f} + f_b\tau_{D\,b}$, and $\tau_{D\,f}$ and $\tau_{D\,b}$ are the relaxation times of the "free" and "bound" water molecules. In the dilute limit $f_b \approx 0$, so that $\tau_D = \tau_{D\,f}$, which is the binary collision time of water molecules in dilute vapors. Librational motions of the water molecules were assumed to cause the braking of the relaxation, hence $\tau_{D\,b} = \tau_{D\,f} + \langle\tau_{lib}\rangle \exp (\Delta_{hb}H/RT)$, where $\langle\tau_{lib}\rangle$ is the inverse of the mean librational frequency of the water molecules and $\Delta_{hb}H$ is the enthalpy for the formation of a hydrogen bond. The enthalpy was taken as $\Delta_{hb}H = -10.6 \pm 0.4\,kJ \cdot mol^{-1}$ and $\langle\tau_{lib}\rangle$ as 0.067 ps, leading to values of the fraction of hydrogen-bonded water molecules, f_b. The resulting values fall between the NMR values of Matubayashi

TABLE 3.2 Dielectric Relaxation Times, τ_D/ps in SCW (Interpolated from Tabulated Data in Ref. [84])

ρ/kg·m^{-3}	$t/°C$		
	400	500	600
100	2.03	1.90	1.77
200	1.03	0.92	0.80
300	0.68	0.63	0.55
400	0.56	0.50	0.44
500	0.52	0.42	
600	0.50		
700	0.47		

et al. [77] or Hoffmann and Conradi [3] (Section 3.3.3) and those obtained from neutron diffraction of Jedlovszky et al. [34] (Section 3.1.2).

Molecular dynamics computer simulations were applied to the dielectric relaxation in SCW by Skaf and Laria [48] and by Guardia and Marti [50, 51]. The former authors used the SPC/E model of water and studied mainly the near-critical states (650 K and densities of 200–700 kg·m^{-3}). They obtained relaxation times of $0.39 \le \tau_D \le 0.44$ ps, in agreement with the experimental values of Okada et al. [84, 86] only at the higher densities, $\rho > \rho_c$. In their first paper Guardia and Marti [50] limited the discussion to the near critical region (650 K) and again used the SPC/E model of water and found relaxation times approaching from below the experimental values as the densities approach 700 kg·m^{-3}, but being much smaller than the experimental values at $\rho \le \rho_c$. In their later paper [51] they also examined the relaxation times in SCW at 720 and 870 K (nominally 447 and 597°C) and again found discrepancies (too low values of τ_D) at densities $\le$700 kg·m^{-3}.

3.4 THE EXTENT OF HYDROGEN BONDING IN SCW

Several aspects of the extent of hydrogen bonding in SCW have been discussed above, according to the methods used for obtaining this information. The main quantities of interest are the average numbers of hydrogen bonds per any water molecule, χ (a maximum of 4) or the average numbers of hydrogen bonds per water molecules in the system, $\langle n_{HB} \rangle$ (a maximum of 2). The fractions f_i of water molecules with $i = 0$, 1, 2, 3, and 4 hydrogen bonds are a further indication of the extent of hydrogen bonding in SCW. The sizes of clusters of hydrogen-bonded water molecules is still another indication of this. A further point to be considered here is the position (T, ρ) of the percolation threshold, beyond which no connected hydrogen-bonded chains exist anymore, but only small clusters (down to dimers) of water molecules.

The fraction f_0 of monomeric, nonhydrogen-bonded, water molecules in SCW was deduced from its *PVT* properties by means of several scaling equations of state, Section 2.1.3. Gupta et al. [87], derived an expression, based on the Panayiotou–Sanchez lattice fluid hydrogen bonding (LFHB) model [88], which relates f_0 to the temperature and the pressure by means of six specified parameters, three of them scaling the temperature, pressure, and density and three relating to the hydrogen bonding energy, entropy, and volume change, Eqs. (2.7)–(2.10). Smits et al. [89] preferred the associated perturbed anisotropic chain theory (APACT) model of Ikonomou and Donohue [96] with three sites of hydrogen bonding per water molecule [90]. Their expression required four molecular parameters: a characteristic energy and a characteristic volume as well as the enthalpy $\Delta H°$ and entropy $\Delta S°$ for hydrogen bonding, Eq. (2.11). They compared values of f_0 from their approach with values from the earlier LFHB approach of Gupta et al. [87] and the statistical associated fluid theory (SAFT) of Huang and Radosz [91], finding considerable differences. More recently Vlachou et al. [92] developed the LFHB approach [87] and used different values for the hydrogen bonding energy and entropy parameters and a temperature and pressure dependent volume of hydrogen bonding. They obtained a still different set of the fraction of monomeric water f_0. The values of f_0 of SCW obtained according to these models at two characteristic pressures, $P = 50$ and 100 MPa, interpolated and read from the reported figures, are shown in Table 3.3, where considerable differences in the estimated values are apparent. All the approaches agree, of course, that f_0 diminishes with lowered temperatures and increasing pressure (and density).

Apart from the equations of state, values of f_0 were also estimated from other approaches. Marcus [93] derived f_0 values from the PVT data (see below) and these are also shown in Table 3.3. They tend to agree with those of the LFHB model of Gupta et al. [87] at the lower pressure and densities at temperatures $\geq 450°C$ and at the higher pressure more nearly with those of Vlachou et al. [92].

Very recently Tassaing et al. [55] measured the infrared spectra of water and deconvoluted the curves into the contributions from species with zero, one, and two hydrogen bonds. However, the data pertained mostly to water vapors

TABLE 3.3   The Fraction of Water Molecules as Monomers (Nonhydrogen-Bonded), f_0, Obtained from Various Equations of State and PVT Data*

$t/°C$	374	400	450	500	550	600
			$P = 50$ MPa			
$\rho/\text{kg·m}^{-3}$	637	582	404	247	186	151
[87]		0.60	0.67	0.75	0.82	0.85
[89] APACT	0.17	0.19	0.28	0.42	0.52	0.58
[89] SAFT	0.48				0.91	
[92]		0.08	0.28	0.51	0.63	0.72
[93]	0.03	0.33	0.62	0.73	0.80	0.84
[54]		0.30				
			$P = 100$ MPa			
$\rho/\text{kg·m}^{-3}$	721	688	612	521	430	350
[87]		0.56	0.61	0.66	0.70	0.74
[89] APACT	0.14	0.17	0.21	0.26	0.32	0.38
[89] SAFT	0.40				0.77	
[92]		0.01	0.13	0.27	0.39	0.52
[93]	0.00	0.04	0.21	0.37	0.47	0.55

*Note that the quoted densities differ somewhat from those in Table 2.1.

at densities below the critical, and only two data points, for 380°C at densities of 325 and 425 kg·m^{-3} pertained to SCW, with $f_0 = 0.69$ and 0.63. These values are comparable with those derived from SAFT theory [91] in Table 3.3 (at 374°C) on linear extrapolation from the higher densities. They also agree with values from molecular dynamics calculations of Kalinichev and Churakov [40]. From this standpoint, the higher values of f_0 obtained by the LFHB [87] and SAFT [91] theories appear to lead to more reliable values than those from the APACT [89] theory or the modified LFHB theory [92]. The latter two approaches appeared to overestimate the extent of hydrogen bonding in SCW.

The number of hydrogen bonds per water molecule χ and $\langle n_{HB} \rangle$ in SCW has been reported by several authors, based on a variety of methods. As expected, the values of $\langle n_{HB} \rangle$ increase with increasing densities and diminish with increasing temperatures, the values of $\langle n_{HB} \rangle$, interpolated from the reported data, are shown in Table 3.1.

The values χ deduced from the pair correlation functions obtained by isobaric (100 MPa) X-ray diffraction by Gorbaty and Kalinichev at three temperatures [1, 11], summarized by Eq. (3.6), are much smaller than values obtained by other methods and need to be discounted. The hydrogen bonding relative to ambient water, η, reported by Hoffmann and Conradi [3] from the NMR chemical shifts δ against benzene internal standard must be multiplied by $1.73 = \langle n_{HB} \rangle$ for ambient water [3] to yield the $\langle n_{HB} \rangle$ for SCW. They are smaller than most other values and pertain to rather low densities. The δ value taken for dilute water vapor, where $\langle n_{HB} \rangle = 0$, was −6.6 ppm relative to dilute benzene, but that for ambient water, −2.5 ppm, was taken from other authors and other internal standards. If $\delta = -2.8$ for ambient water relative to internal benzene standard could have been used instead (but the solubility of benzene is too low for measurements), a modification of Eq. (3.9), namely, $\langle n_{HB} \rangle = 1.73\eta = 2.785 + 0.391\sigma$ would result, yielding $\langle n_{HB} \rangle$ values commensurate with most others, a procedure that is recommended.

Another NMR study, that of Matsubayasi et al. [77] at 400°C yielded unreasonably large $\langle n_{HB} \rangle$ values, ranging from 0.6 to 1.5 for densities from 190 to 660 kg·m^{-3} (and a $d(\text{O} \cdots \text{H})$ cutoff of 0.23 nm). The cause for these much too large values is their being anchored at values for 300°C reported by Soper et al. [17] from neutron diffraction with isotope substitution, namely 1.68 or 1.80 at 720 kg·m^{-3} (9.5–10 MPa). The latter authors also reported the value $\langle n_{HB} \rangle = 1.50$ at 400°C and 658 kg·m^{-3} (80 MPa), much larger than values obtained by others.

On the other hand, the values of $\langle n_{HB} \rangle$ obtained by computer simulations (MC by Kalinichev and Bass [30] and MD by Mountain [36] and Yoshii et al. [45]) agree much better with each other and also with values deduced from the IR absorption data of Franck and Roth [53] and calculated from the equations of state (EoS) by Gupta et al. [94G]. The MD data of Yoshii et al. [45] were reported for the nominal temperature of $T = 600$ K, but since the critical temperature for the model used was $T_c = 561$ K, this means a reduced temperature of $T_r = T/T_c = 1.07$, that corresponds to 430°C. The integrated areas B below the O–D stretch peak of dilute HOD in supercritical H_2O at 400°C reported by Franck and Roth [53] were divided by the area at 30°C to obtain the relative amount of hydrogen bonding and were multiplied by 1.73 to yield $\langle n_{HB} \rangle$ shown in Table 3.3.

On a quite different basis Marcus [93] calculated values of $\langle n_{HB} \rangle$ in SCW primarily from the PVT data at 400–600°C (35–100 MPa, 200–700 kg·m^{-3}) on the basis of the model of Lamanna et al. [94], applied by them at $\leq$230°C. The semiempirical model recognizes the existence of water molecules with zero and one hydrogen bonds (monomers and dimers) as well as species (trimers) with two hydrogen bonds. Species with three (cyclic trimers or tetramers) or four hydrogen bonds are presumed to be either so rare that they can be neglected or that their effects can be subsumed into those of species with two hydrogen bonds. The fractions of such species are given by $f_0 + f_1 + f_{\geq 2} = 1$. The probability p of the formation of a hydrogen bond between two water molecules is taken to be independent of any already existing hydrogen bonds, that is, cooperative hydrogen bonding in SCW is negated [94]. The fractions of the species are then given by

$$f_0 = 1/[1 + 4p/(1-p) + 6p^2/(1-p)^2]$$
$$f_1 = 4pf_0/(1-p) \tag{3.12}$$
$$f_2 = 6p^2f_0/(1-p)^2$$

The probability of hydrogen bonding had to be modified from that employed by Lamanna et al. [94] at lower temperatures, to recognize the pressure dependence of the energy parameter, to yield:

$$p = 1 - \exp[-E(P)/R(T-T_0)] \tag{3.13}$$

The molar volume of the SCW is a weighted sum of the volumes of the species:

$$V(T,P) = f_0 V_0 + f_1 V_1 + f_2 V_2 \tag{3.14}$$

neglecting for now species with more than two hydrogen bonds. The volumes of the species, in turn, are temperature- and pressure-dependent as

$$V_i = V_{i0}(P)[1 + \alpha_i(P)(T-T_0)^2] \quad (i = 0, 1, 2) \tag{3.15}$$

To limit the number of fitting parameters required, the molar volume of the trimer (with two hydrogen bonds) was taken as $^3/_2$ that of the dimer (with one hydrogen bond) and their expansibilities were taken to be equal. The following parameters were then required to fit the *PVT* data (Section 2.1.1.) to Eqs. (3.12)–(3.15): $T_0 = 637$ K, $E(P)/\text{J}\cdot\text{mol}^{-1} = 18 + 0.028\,(P/\text{MPa})^2$, $V_{00}/\text{cm}^3\cdot\text{mol}^{-1} = 27.1 - 0.27\,(P/\text{MPa})$, $V_{10}/\text{cm}^3\cdot\text{mol}^{-1} = 29.6 - 0.072\,(P/\text{MPa})$, $\alpha_0/\text{K}^{-2} = -1.7 \times 10^{-5} + 0.44\,(P/\text{MPa})^{-2}$, and $\alpha_1/\text{K}^{-2} = -0.20 \times 10^{-5} + 0.55\,(P/\text{MPa})^{-2}$.

The fractions $f_i(T, P)$ depend only on T_0 and the expression for $E(P)$, requiring three parameters, via the probabilities p according to Eqs. (3.12) and (3.13) and can, therefore be presented in a three-dimensional grid as in Fig. 3.4.

Although derived from fitting the extensive *PVT* data, the same fractions should be applicable to other measurements, and this was indeed done [93]. An expression analogous to Eq. (3.14) was used with these $f_i(T, P)$ to fit the constant volume heat capacities, C_v. The C_v values (Section 2.2.1) for SCW are appreciably larger than the well known ones for monomeric water, $C_{v0} = 21.79 + 0.0158(t/^\circ C)\,\text{J}\cdot(\text{K}^{-1}\cdot\text{mol}^{-1})$, due to the larger moments of inertia for the rotation of the dimers and trimers and the vibrations of the hydrogen bonds. Fairly good fits were obtained with the above C_{v0}, and the values for the dimer: $C_{v1} = 65.73 + 0.0350(t/^\circ C)\,\text{J}\cdot(\text{K}^{-1}\cdot\text{mol}^{-1})$ (obtained from spectroscopy) and of the single fitting parameter for the linear trimer: $C_{v2} = 2.3 \times 10^7\,(P/\text{MPa})^{-3}$. Self-diffusion coefficients (Section 2.4.2) could also be fitted with an expression analogous to Eq. (3.14) and the $f_i(T, P)$ derived from the *PVT* data. Three fitting parameters were required: $D_0 = 157$, $D_1 = 11$, and

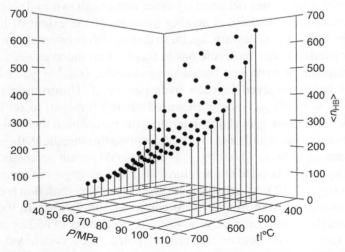

FIGURE 3.4   The average number of hydrogen bonds per water molecule in SCW [93].

$D_2 = 49$, all in $10^{-9}\,\mathrm{m}^2\cdot\mathrm{s}^{-1}$. That the value of D_2 is larger than that of D_1 is attributed to more efficient diffusion of the trimer by rotation. Fair agreement is also achieved with the values of $1 - f_0$ deduced from the NMR chemical shifts (Section 3.3.3 and above) [93].

The model and the calculated $f_i(T, P)$ were applied also to the Kirkwood dipole orientation parameter g, Eqs. (1.11) and (1.12) (Section 1.4), that could be calculated for SCW from permittivity data (Section 2.3.1) [95]. The equation analogous to Eq. (3.14) is then

$$g = 1 + f_0 N(T,P)\cos\langle\theta_0\rangle + f_1 N(T,P)\cos\langle\theta_1\rangle + f_2 N(T,P)\cos\langle\theta_2\rangle \quad (3.16)$$

where $N(T, P)$ is the number of nearest neighbors and $\langle\theta\rangle$ is the average angle between the dipoles of neighboring water molecules. For nonhydrogen-bonded molecules the average random angle is $90°$, hence $\cos\langle\theta_0\rangle = 0$ and the term involving such molecules vanishes. There are two probable configurations for singly hydrogen-bonded water molecules: free rotation around this bond or minimal oxygen lone pair repulsion, leading to a choice for $\cos\langle\theta_1\rangle$ between 0.49 and 0.39. The estimation of the average angle for the trimers with two hydrogen bonds is difficult and was not attempted, so that calculations were limited to temperatures and pressures where f_2 is negligible. General agreement was found [93] between $1 + f_1 N(T, P)\cos\langle\theta_1\rangle$ and values of g calculated from the permittivities, but the latter themselves cover a range of values depending on uncertainties in $\varepsilon(\infty)$ (Section 2.3.1) [95].

The values of $\langle n_{HB}\rangle = f_1 + 2f_2$ obtained in this manner are shown in Fig. 3.4 and those of f_0, the fraction of nonhydrogen-bonded water molecules in SCW arising from the model just discussed are shown in Fig. 3.5. They can be compared with values obtained by other authors shown in Table 3.3.

The percolation threshold is another measure of the extent of hydrogen bonding in SCW. It is necessary, again, to distinguish between the number of hydrogen bonds a water molecule has in the fluid on the average, χ, and the average number of hydrogen bonds per molecule, $\langle n_{HB}\rangle = \chi/2$, two water molecules being involved in each hydrogen bond. Unfortunately, in the literature the symbol n_{HB} is sometimes used where χ is meant, so, for instance, in the paper of Blumberg et al. [5] specifying the percolation threshold as 1.53 (later rounded by other authors to 1.55). Accepting this threshold, then SCW in thermodynamic states at which $\langle n_{HB}\rangle \geq 0.77$ should permit percolation, since then their hydrogen-bonded clusters have the required degree of connectivity.

Jedlovszky et al. [23, 34, 35] discussed *inter alia* the percolation probability in SCW, as arising from Monte Carlo computer simulations. In the first of these papers they considered SCW at 400°C and a density of $652\,\mathrm{kg\cdot m^{-3}}$ only and also several states below the critical point. They concluded that the required connectivity for percolation is lost at the critical point [34]. In their

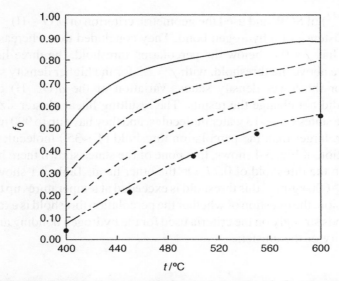

FIGURE 3.5 The fraction of monomeric (nonhydrogen-bonded) water molecules in SCW, curves calculated from Ref. [93]: ——— at 40 MPa, – – – – at 60 MPa, - - - - - - at 80 MPa, and –··–··–··– at 100 MPa. From Vlachou et al. [92] at 100 MPa: •.

subsequent papers they attempted to locate the percolation threshold line in SCW more precisely, as presented in Section 3.2.1: within states of 384, 430, 480, and 530°C at supercritical densities (368–428 kg·m^{-3}) [23], at 430°C at densities of 329–479 kg·m^{-3} [35] and 200–500 kg·m^{-3} read from figures [07Pa]. (The temperatures were adjusted from the nominal ones in view of the $T_c = 367$°C of the SPC/E model used.) The criteria for percolation were that ≥50% of the configurations in their simulations were of "infinite" clusters, spanning the simulation box from boundary to boundary in at least one dimension or alternatively that the distribution of clusters of n molecules was $p(n) \sim n^{-2.19}$. The results, in turn, devolved on whether a geometric criterion for the hydrogen bond ($d(\mathrm{O-O}) \leq 0.35$ nm and $d(\mathrm{O} \cdots \mathrm{H}) \leq 0.25$ nm) or an energetic criterion ($e_{\mathrm{HB}} < -11.5$ kJ·mol^{-1}) were used. The latter criterion is more strict and permitted percolation only in the lowest temperature studies in Ref. [23], namely just above the critical point, whereas the former, more lenient geometric criterion permitted percolation also at the next two higher temperatures, that is, 430 and 480°C. At 430°C the threshold appeared to be a density ≥410 kg·m^{-3} [35, 100].

Bernabei et al. [21, 22] made new neutron diffraction measurements with isotope substitution (NDIS: H_2O, equimolar $H_2O + D_2O$, and D_2O) and analyzed them by means of the empirical potential structure refinement method. They applied this to one gas-like state (400°C and 116 kg·m^{-3}) and three liquid-like states (400°C and 580 and 750 kg·m^{-3} and 480°C and

750 kg·m^{-3}) of SCW and used the geometric criterion of $d(\text{O} \cdots \text{H}) \leq 0.24$ nm for the existence of a hydrogen bond. They concluded that whereas the gas-like state has $\chi = 0.8$, below the percolation threshold, the three liquid-like states were above the threshold, with $\chi = 2.2$ for the higher density states and $\chi = 1.9$ for the lower density states. Variation of the $d(\text{O} \cdots \text{H})$ cutoff by ± 0.1 nm did not change the results. The resulting linear cluster sizes of the liquid-like states was ~ 13 water molecules, and they have up to 900 molecules altogether, larger than the percolation threshold of ~ 350 molecules.

Inspection of Fig. 3.4 shows, that none of the states shown there has $\langle n_{\text{HB}} \rangle$ larger than the threshold of 0.77. On the other hand, Table 3.1 shows that at densities ≥ 600 kg·m^{-3} this threshold is exceeded at temperatures up to 600°C. In conclusion, the question of whether the percolation threshold is exceeded or not, depends strongly on the criteria used for the hydrogen bonding and for the connectivity of the clusters.

3.5 THE DYNAMICS OF WATER MOLECULES IN SCW

Certain spectroscopic measurements and molecular dynamics computer simulations provide information regarding the dynamics of individual water molecules and of the hydrogen bonding between them in SCW. For results obtained by MD, the original reported temperatures were adjusted according to the critical temperatures resulting for the model employed, see Section 3.2.2.

Just above the critical point, 380°C, the hydrogen bond lifetime was studied by Tassaing et al. [24], using incoherent neutron scattering. The lifetime remained nearly constant near 0.19 ps up to a density of 600 kg·m^{-3} and decreased then to ~ 0.13 ps at 900 kg·m^{-3}, reflecting the higher incidence of collisions at the higher densities. Skarmoutsos and Guardia [57] used MD simulations in SCW at 385°C and 61–644 kg·m^{-3} to calculate the lifetimes of hydrogen bonds, defined geometrically: $d(\text{O}-\text{O}) \leq 0.36$ nm and $d(\text{O} \cdots \text{H}) \leq 0.24$ nm. The continuous lifetime, τ_{HBc}, was 0.075 ps over the density range and the intermittent lifetimes, τ_{HBi}, decrease from 0.62 ps at 61 kg·m^{-3}, to a minimum of 0.44 ps near the critical density, and increase slightly to 0.47 ps at 644 kg·m^{-3}.

Skarmoutsos and Samios [56] studied the density rearrangement times, $\tau_{\Delta\rho}$, in SCW at 385°C and 61–644 kg·m^{-3} by MD simulations. These times decreased from ~ 1.3 ps at the lowest to ~ 0.3 ps at the highest density examined.

The rotational relaxation time, τ_{rot}, was derived from the MD simulations of Yoshii et al. [45] using a polarizable SPC model at 400°C (after temperature adjustment). It was minimal, $\tau_{\text{rot}} \sim 0.04$ ps, at a density of 147 kg·m^{-3} and it increases to $\tau_{\text{rot}} \sim 0.10$ ps at 1000 kg·m^{-3}. Matubayashi et al. [49] too studied the rotational dynamics of SCW at 400°C and found reorientational correlation times, τ_{rot}, increasing from 0.048 ps at 100 kg·m^{-3} to 0.064 ps at 600 kg·m^{-3}, but depending on the assumed quadrupole coupling constant (varying between 256 and 308 kHz).

Guardia and Marti [50, 51] used MD simulations of SCW reorientational dynamics at 380°C at a wide range of densities, from 10 to 700 kg·m^{-3}. They examined the orientational correlation times along four vectors. Along the dipole vector τ_1 increased steadily from 0.027 to 0.225 ps, whereas τ_2 showed a minimum of 0.031 ps at 50 kg·m^{-3}. Along the O–H bond and the line connecting the two hydrogen atoms the times also showed minima at this density, and in a direction perpendicular to the plane of the molecule the minimum occurred at 100 kg·m^{-3}. At the highest density, 700 kg·m^{-3}, the times were ~0.20 ps for τ_1 and 0.08 ps for τ_2 at all four directions.

Jonas et al. [75] applied (proton) NMR to SCW at 400–600°C and fairly low densities, 50–350 kg·m^{-3}. They reported the spin-lattice relaxation times T_1 that increased with increasing densities but diminished with increasing temperatures, but were unable to obtain the angular momentum correlation times τ_J from their data. Matubayashi et al. [49] measured the spin-lattice relaxation times T_1 for D_2O at 400°C at densities of 0.1–0.6 relative to those of the ambient liquid. In this case the relaxation is governed by the quadrupolar relaxation mechanism, providing a second order reorientational time τ_{2R}. The derived values of τ_{2R} ranged from 0.047 to 0.082 ps, increasing with the density. Yoshida et al. [79, 80] used high-resolution proton NMR to study SCW at 400°C. The solvation relaxation time, τ_S, is the relaxation time in the solvation shell of a given water molecule. It declines with the number n of molecules in the clusters, from 0.04 ps at $n = 4$–0.01 ps at $n = 22$ at the supercritical density of 600 kg·m^{-3}.

Okada et al. [84] measured dielectric relaxation in SCW up to ~600°C and densities ≤700 kg·m^{-3}, yielding the Debye relaxation times, τ_D, shown in Table 3.2. Okada et al. [84] and Yao and Hiejima [85] distinguished between τ_{Df} and τ_{Db}, the relaxation times of the "free" and "bound" water molecules, as dealt with in Section 3.3.4. Skaf and Laria [48] studied the dielectric relaxation times, τ_D, by means of MD simulations as did Guardia and Marti [50], again, as dealt with in Section 3.3.4.

In summary, the continuous lifetime of a hydrogen bond in SCW is of the order of 0.1 ps, decreasing somewhat with increasing densities as the collision rate increases. The intermittent lifetime is about fivefold longer. The orientational relaxation times are shorter, of the order of 0.05 ps, but depend on the vectors around which the rotation is considered. The Debye (molecular cooperative) relaxation times are considerably larger than such values, above 0.4 ps, but there exists disagreement between the experimental (microwave) and computer simulated (MD) results at densities ≤ 400 kg·m^{-3}.

3.6 SUMMARY

The structure of SCW is expressed by the pair correlation function, $g(r)$ in Eq. (3.1) that describes the probability of finding another water molecule at a distance r from a given one. The mutual arrangement of two neighboring water molecules is shown in Fig. 3.1. X-ray diffraction (Section 3.1.1), arising from the electronic shell of an atom, describes only the O–O distances at which water molecules are to be found in terms of the structure factors $S(I)$ and Eqs. (3.2)–(3.4). Neutron diffraction (Section 3.1.2), arising from atomic nuclei, in particular modified by the isotope substitution method (NDIS), provides the partial pair correlation functions: $g(O–O, r)$, $g(O–H, r)$, and $g(H–H, r)$ (Fig. 3.2). A controversy that arose in the 1990s from the interpretation of neutron diffraction results concerning whether any HBs at all exist in SCW was subsequently resolved in confirmation of their existence and its extent. The $g(O–H, r)$ resulting from these measurements did show a distinct small peak at 0.19 nm that is considered as the "signature" of the existence of an HB.

Certain more general criteria are invoked in order to decide whether an HB does or does not exist, as described in the text. The diffraction experiments provided the mean number of hydrogen bonds (HBs) per water molecule, $\langle n_{HB} \rangle$. The average number of hydrogen bonds that a given water molecule has, χ, is twice $\langle n_{HB} \rangle$, because two water molecules are involved in each HB. The quantity $f_4 = \chi/4$, the fraction of tetrahedral hydrogen bonding persisting in high temperature water was obtained from X-ray diffraction, decreasing linearly with the temperature up to 500°C and densities between 700 and 1100 kg·m^{-3}, Eq. (3.6). Recent neutron diffraction data of SCW, depending on the density, whether it is liquid-like (750 kg·m^{-3} and 400 and 480°C) or gas-like (116 kg·m^{-3} and 400°C), were interpreted in terms of the presence or absence of percolation, that is, some hydrogen bonding network. The χ of a water molecule in the gas-like state is only 0.8, below the percolation threshold of $\chi = 1.55$, which in turn is exceeded in the liquid-like states, where it is 2 on the average and $f_4 = 5–10\%$.

MC computer simulations (Section 3.2.1) of SCW in the 1990s are connected with Kalinichev et al., who used the empirical TIP4P potential on SCW at 400–1000°C at densities ranging from 30.5 to 1666 kg·m^{-3}. Partial pair correlation functions: $g(O–O, r)$, $g(O–H, r)$, and $g(H–H, r)$ were shown, exhibiting the hydrogen bond signature peak at 0.19 nm in $g(O–H, r)$, the absence of the tetrahedral structural feature at 0.45 nm in $g(O–O, r)$, and hardly any rotational orientation correlation in $g(H–H, r)$. The findings were augmented with criteria for HB formation, leading to values of χ increasing from 0.2 at 10 MPa (30.5 kg·m^{-3}) to 2.5 at 10 GPa (1666 kg·m^{-3}). The percolation threshold ($\chi = 1.55$) is exceeded only for temperatures below 600°C at any of the available densities. The $\chi = f(T, P, \rho)$ values correlate well with f_0, the fraction of monomeric water molecules (those devoid of any HB), Eq. (3.7), the latter can be compared with values obtained from the equation of state (Section 2.1.3) and other sources (Section 3.4). More recently Jedlovszkyal. concentrated on the HB percolation properties of SCW at 430–534°C at both gas-like (17–180 kg·m^{-3}) and liquid-like densities (180–428 kg·m^{-3}) using the SPC/E potential function. They confirmed that percolation occurs at densities ≥ 344 kg·m^{-3}.

In contrast with MC simulations, for which the temperature is an input quantity, for molecular dynamics simulations (Section 3.2) it is an output, and the derived virtual critical temperature T_c generally does not agree with the experimental one of SCW, 374°C. The potentials used in various water models lead to values $301 \leq T_{c\ model} \leq 437$, and it is expedient to report the reduced temperature $T_r = T/T_{c\ model}$, rather than the reported nominal temperatures T. Mizan et al. discussed critically (Section 3.2.2) the choice of the water model potential function and showed that rigid models, such as SPC or TIP4P, yield critical points far from the experimental one, whereas flexible models succeed much better in reproducing the thermodynamic properties, but there are still variations among these models. For $T_r = 1.34$ and densities of 115 (gas-like), 257–659 (liquid-like) kg·m^{-3} the 0.19 nm peak was clearly shown by the $g(O–H, r)$ curves of all the SCW states, but its intensity diminished with increasing densities. Self-diffusion coefficients calculated from the simulations agreed well with experimental values (Section 2.4.2). Many other authors have applied MD simulations to SCW using various potential models, numbers of molecules involves, and over wide states of temperatures and pressures densities). The resulting $\langle n_{HB} \rangle = \chi/2$ values are summarized in Table 3.1, from which percolation thresholds $\langle n_{HB} \rangle = 1.55/2 \sim 0.77$ can be deduced, and compared with statements of the various authors. Local density inhomogeneities were studied in SCW at $T_r = 1.03$ over a wide density range: 61–644 kg·m^{-3} by MD simulations. First shell density enhancement of 43% and second shell enhancement of 21% were found at a density of 122 kg·m^{-3}, the relative quantities diminishing at increasing densities. Local densities were more relevant to the hydrogen bond connectivity than the bulk density.

Infrared spectroscopy of SCW (Section 3.3.1) was most profitably conducted with dilute HOD in H_2O or in D_2O. The O–H (O–D) stretching frequency increased (a blue shift) as the density decreased and the O–O distance increased, leading to diminished χ. Raman spectroscopy (Section 3.3.2) was more widely applied to SCW and to produced more meaningful results than IR measurements. Again dilute HOD in SCW was used, and the favored interpretation of the results was in terms of increasing fractions of non- or weakly hydrogen-bonded water molecules with O–H stretching frequencies ≥ 3550 cm^{-1} as the temperature is increased and/or the density is decreased. The 3250 cm^{-1} shoulder of the Raman band was assigned by Walrafen to correlated collective motions of hydrogen-bonded molecules and the $3400\,\mathrm{cm}^{-1}$ peak to uncorrelated stretching vibrations of the O–H bonds. Over a wide temperature range, including SCW, the intensity ratio, $R_{c/u}$, of the correlated to uncorrelated bands is linear with the structural correlation lengths, suggesting that hydrogen-bonded clusters are present there that have some degree of collective motions of their molecules. Figure 3.3 drawn following Walrafen et al. illustrates the frequency shifts near the critical point. As the temperature was raised a blue shift was seen, indicating the breaking of hydrogen bonds. The Raman O–H stretching frequencies of SCW were red-shifted linearly at a given temperature as the density increased from 200 to $600\,\mathrm{kg\cdot m}^{-1}$, as a result of hydrogen bonding.

X-ray Raman scattering at the oxygen K-edge, rather than light Raman scattering, was applied to SCW at $380\,°\mathrm{C}$ and $540\,\mathrm{kg\cdot m}^{-3}$. The intensity of the post-edge region, an energy region absent in water vapor, points to a large degree of hydrogen bonding. Clusters of 5–10 water molecules represent the hydrogen-bonded domains in the just-critical water studied. The fraction of water molecules with two nonhydrogen-bonded O–H groups, f_0, was estimated from the pre-edge features as $\sim 35\%$.

The proton NMR chemical shifts in SCW (Section 3.3.3) at $400–600\,°\mathrm{C}$ were measured, using dilute benzene as an internal reference. The water resonance shifts to lower frequencies at increasing temperatures where the hydrogen bond network is increasingly destroyed. A linear relationship between the extent of hydrogen bonding η relative to ambient water (for which $\chi = 3.46$) and the chemical shift δ resulted, Eq. (3.9). Chemical shifts δ for $^{17}\mathrm{O}$ were also linear with η. The proton chemical shift of SCW was calculated theoretically, leading to 80% monomer and 20% dimer at the critical temperature and density. No cooperativity of hydrogen bonding needs to be taken into account for SCW.

In view of all these studies, as well as some information, f_0, from the EoSs the extent of the hydrogen bonding in SCW is discussed in Section 3.4. Consensus exists only weakly, but the existence of some hydrogen bonding in SCW is firmly established. How much of it occurs at various thermodynamic states depends on the criteria used for defining an HB. The values of f_0 of SCW obtained according to theoretical models (Section 2.1.3) from PVT data at two characteristic pressures, $P = 50$ and $100\,\mathrm{MPa}$, are shown in Table 3.3, where considerable differences in the estimated values are apparent. All the approaches agree, of course, that f_0 diminishes with lowered temperatures and increasing pressure (and density). The number of hydrogen bonds per water molecule are shown in Table 3.1, as already mentioned. Figure 3.4 shows $\langle n_{\mathrm{HB}} \rangle = f_1 + 2f_2$ obtained, Eq. (3.12), from the PVT data. The constant volume heat capacity C_v (Section 2.2.1), the self-diffusion coefficient D (Section 2.4.2), and the Kirkwood dipole orientation parameter g (Section 1.4), that could be calculated for SCW from permittivity data (Section 2.3.1) could also be fitted with the calculated f_0, f_1, and f_2. Values of f_0, the fraction of nonhydrogen-bonded water molecules in SCW are shown in Fig. 3.5. Fair agreement is also achieved with the values of $1 - f_0$ deduced from the NMR chemical shifts (Section 3.3.3).

The dynamics of the water molecules and HBs in SCW was derived from incoherent neutron scattering, several spectroscopic studies, and MD computer simulations. The former method led to the conclusion that orientational correlation between neighboring water molecules is almost entirely lost. The HB lifetime remained nearly constant near 0.19 ps up to a density of $600 \, \text{kg} \cdot \text{m}^{-3}$ and decreased then to $\sim 0.13 \, \text{ps}$ at $900 \, \text{kg} \cdot \text{m}^{-3}$, reflecting the higher incidence of collisions at the higher densities. NMR spin-lattice relaxation times T_1 yielded angular momentum correlation times τ_J increasing with the temperature at a constant density $(350 \, \text{kg} \cdot \text{m}^{-3})$ from 0.0641 ps at 400°C to 0.103 ps at 700°C, but decreasing with the density at a constant temperature. The reorientation correlation times τ_θ were somewhat lower than τ_J, and were in the range 0.008–0.060 ps. Dielectric relaxation in SCW up to $\sim 600°C$ and densities $\leq 700 \, \text{kg} \cdot \text{m}^{-3}$ yielded the Debye (cooperative) relaxation times, τ_D, shown in Table 3.2. The relaxation times of the "free" and "bound" water molecules, $\tau_{D \, f}$ and $\tau_{D \, b}$, were dealt with in Section 3.3.4.

MD computer simulations were applied to the dielectric relaxation in SCW yielding Debye relaxation times of $0.39 \leq \tau_D \leq 0.44$ ps, in agreement with the experimental values only at high densities, $> \rho_c$. The continuous lifetime of HBs defined geometrically, τ_{HBc}, was obtained by MD simulations in SCW at 385°C and 61–$644 \, \text{kg} \cdot \text{m}^{-3}$ as 0.075 ps over the density range. The intermittent lifetimes, τ_{HBi}, decreased from 0.62 ps at $61 \, \text{kg} \cdot \text{m}^{-3}$, to a minimum of 0.44 ps and increased slightly to 0.47 ps at $644 \, \text{kg} \cdot \text{m}^{-3}$. The rotational relaxation time, τ_{rot}, was derived from the MD simulations was minimal, $\tau_{rot} \sim 0.04$ ps, at a density of $147 \, \text{kg} \cdot \text{m}^{-3}$ and it increases to $\tau_{rot} \sim 0.10$ ps at $1000 \, \text{kg} \cdot \text{m}^{-3}$. Thus, when dealing with relaxation times it is imperative to define the process to which they pertain, since they span a wide range, from $\tau_{rot} \sim 0.04$ ps for single molecular rotation and reorientation to $\tau_D = 2$ ps at the lowest temperature and density for molecular cooperative movements.

REFERENCES

1. Yu. E. Gorbaty and R. B. Gupta, *Ind. Eng. Chem. Res.* **37**, 3026 (1998).
2. R. Kumar, J. R. Schmidt, and J. L. Skinner, *J. Chem. Phys.* **126**, 204107 (2007).
3. M. M. Hoffmann and M. S. Conradi, *J. Am. Chem. Soc.* **119**, 3811 (1997).
4. W. Kohl, H. A. Lindner, and E. U. Franck, *Ber. Bunsenges. Phys. Chem.* **95**, 1586 (1991).

5. R. L. Blumberg, H. E. Stanley, A. Geiger, and P. Mausbach, *J. Chem. Phys.* **80**, 5230 (1984).

6. Yu. E. Gorbaty and Yu. N. Dem'yanets, *Zhur. Strukt. Khim.* **23**, 73 (1982).

7. Yu. E. Gorbaty and Yu. N. Dem'yanets, *Zhur. Strukt. Khim.* **24**, 66, 74 (1983).

8. Yu. E. Gorbaty and Yu. N. Dem'yanets, *Chem. Phys. Lett.* **100**, 450 (1983).

9. K. Yamanaka, T. Yamaguchi, and H. Wakita, *J. Chem. Phys.* **101**, 9830 (1994).

10. A. V. Okhulkov, Yu. N. Dem'yanets, and Yu. E. Gorbaty, *J. Chem. Phys.* **100**, 1578 (1994).

11. Yu. E. Gorbaty and A. G. Kalinichev, *J. Phys. Chem.* **99**, 5336 (1995).

12. H. Ohtaki, T. Radnai, and T. Yamaguchi, *Chem. Soc. Rev.* **26**, 41 (1997).

13. M. Inui, Y. Kajihara, Y. Azumi, K. Matsuda, and K. Tamura, *J. Phys. Conf. Rep.* **215**, 012090 (2010).

14. P. Postorino, R. H. Tromp, M. A. Ricci, A. K. Soper, and G. W. Neilson, *Nature* **366**, 668 (1993).

15. R. H. Tromp, P. Postorino, G. W. Neilson, M. A. Ricci, and A. K. Soper, *J. Chem. Phys.* **101**, 6210 (1994).

16. M.-C. Bellissent-Funel, T. Tassaing, H. Zhao, D. Beysens, B. Guillot, and Y. Guhssani, *J. Chem. Phys.* **107**, 2942 (1997).

17. A. K. Soper, F. Bruni, and M. A. Ricci, *J. Chem. Phys.* **106**, 247 (1997).

18. A. Botti, F. Bruni, M. A. Ricci, and A. K. Soper, *J. Chem. Phys.* **109**, 3180 (1998).

19. T. Tassaing, M.-C. Bellissent-Funel, B. Guillot, and Y. Guhssani, *Europhys. Lett.* **42**, 265 (1998).

20. M.-C. Bellissent-Funel, *J. Mol. Liquids* **90**, 313 (2001).

21. M. Bernabei, A. Botti, F. Bruni, M. A. Ricci, and A. K. Soper, *Phys. Rev. E* **78**, 021505 (2008).

22. M. Bernabi and M. A. Ricci, *J. Phys. Condens. Matt.* **20**, 494208 (2008).

23. L. Pátay and P. Jedlovszky, *J. Chem. Phys.* **123**, 024502 (2005).

24. T. Tassaing, Y. Danten, and M. Besnard, *Pure Appl. Chem.* **76**, 133 (2004).

25. T. Otomo, H. Iwase, Y. Kameda, N. Matubayashi, K. Itoh, S. Ikeda, and M. Nakahara, *J. Phys. Chem. B* **112**, 4687 (2008).

26. S. F. O'Shea and P. R. Tremaine, *J. Phys. Chem.* **84**, 3304 (1980).

27. T. I. Mizan, P. E. Savage, and R. M. Ziff, *J. Phys. Chem.* **98**, 13067 (1994).

28. A. G. Kalinichev and K. Heinzinger, in S. Saxena, Ed., *Advances in Physical Geochemistry*, Ch. 1, Springer, New York, 1992.

29. A. G. Kalinichev, *Z. Naturforsch.* **46a**, 433 (1991).

30. A. G. Kalinichev and J. D. Bass, *Chem. Phys. Lett.* **231**, 301 (1994).

31. A. G. Kalinichev and J. D. Bass, *J. Phys. Chem. A* **101**, 9720 (1997).

32. A. G. Kalinichev and S. V. Churakov, *Fluid Phase Equil.* **183–184**, 271 (2001).

33. N. Matubayashi, C. Wakai, and M. Nakahara, *J. Chem. Phys.* **110**, 8000 (1999).

34. P. Jedlovszky, J. P. Brodholt, F. Bruni, M. A. Ricci, A. K. Soper, and R. Vallauri, *J. Chem. Phys.* **108**, 8528 (1998).

35. L. Pátay, P. Jedlovszky, I. Brovchenko, and A. Oleinikova, *J. Phys. Chem. B* **111**, 7603 (2007).

36. R. D. Mountain, *J. Chem. Phys.* **90**, 1866 (1989).

37. P. T. Cummings, H. D. Cochran, J. M. Simonson, R. E. Mesmer, and S. Karaborni, *J. Chem. Phys.* **94**, 5606 (1991).

38. A. G. Kalinichev, *Ber. Bunsengesel. Phys. Chem.* **97**, 872 (1995).

39. A. G. Kalinichev and K. Heinzinger, *Geochim. Cosmochim. Acta* **59**, 641 (1995).

40. A. G. Kalinichev and S. V. Churakov, *Chem. Phys. Lett.* **302**, 411 (1999).

41. S. V. Churakov and A. G. Kalinichev, *J. Struct. Chem.* **40**, 548 (1999).

42. S. Krishtal, M. Kiselev, Y. Puhovski, T. Kerdcharoen, S. Hanongbua, and K. Heinzinger, *Z. Naturforsch.* **56a**, 579 (2001).

43. L. X. Dang, *J. Phys. Chem. B* **102**, 620 (1998).

44. C. C. Liew, H. Inomata, K. Arai, and S. Saito, *J. Supercrit. Fluids* **13**, 83 (1998).

45. N. Yoshii, H. Yoshie, S. Miura, and S. Okazaki, *J. Chem. Phys.* **109**, 4873 (1998).

46. V. E. Petrenko, M. L. Antipova, O. V. Ved' and A. V. Borovlov, *Struct. Chem.* **18**, 505 (2007).

47. S. Bastea and L. E. Fried, *J. Chem. Phys.* **128**, 174502 (2008).

48. M. S. Skaf and D. Laria, *J. Chem. Phys.* **113**, 3499 (2000).

49. N. Matubayashi, N. Nakao, and N. Nakahara, *J. Chem. Phys.* **114**, 4107 (2001).

50. E. Guardia and J. Marti, *J. Mol. Liquids* **101**, 137 (2002).

51. E. Guardia and J. Marti, *Phys. Rev. E* **69**, 011502 (2004).

52. P. J. Dyer and P. T. Cummings, *J. Chem. Phys.* **125**, 144519 (2006).

53. E. U. Franck and K. Roth, *Disc. Faraday Soc.* **43**, 108 (1967).

54. D. Swiata-Wojcik and J. Szala-Bilnik, *J. Chem. Phys.* **134**, 054121 (2011).

55. T. Tassaing, P. A. Garrain, D. Bégué, and I. Baraille, *J. Chem. Phys.* **133**, 0324103 (2010).

56. I. Skarmoutsos and J. Samios, *J. Phys. Chem. B* **110**, 21931 (2006).

57. I. Skarmoutsos and E. Guardia, *J. Chem. Phys.* **132**, 074502 (2010).

58. A. Kandratsenka, D. Schwarzer, and P. Vöhringer, *J. Chem. Phys.* **128**, 244510 (2008).

59. W. A. P. Luck, *Ber. Bunsenges. Phys. Chem.* **60**, 626 (1965).

60. W. A. P. Luck, *Disc. Faraday Soc.* **43**, 115 (1967).

61. W. A. P. Luck and W. Ditter, *Z. Naturforsch.* **24b**, 482 (1969).

62. Yu. E. Gorbaty, *Problems of Physical-Chemical Petrology*, Vol. II, p. 15, Nauka, Moscow, 1979.

63. G. V. Bondarenko, Yu. E. Gorbaty, A. V. Okhulkov, and A. G. Kalinichev, *J. Phys. Chem. A* **110**, 4042 (2006).

64. Yu. E. Gorbaty and G. V. Bondarenko, *Appl. Spectrosc.* **53**, 908 (1999).

65. J. D. Frantz, J. Dubessy, and B. Mysen, *Chem. Geol.* **106**, 9 (1993).

66. G. E. Walrafen and T. C. Chu, *J. Phys. Chem.* **99**, 11225 (1995).

67. D. A. Masten, B. R. Foy, D. M. Harradine, and R. B. Dyer, *J. Phys. Chem.* **97**, 8557 (1993).

68. Y. Ikushima, K. Hatakeda, and N. Saito, *J. Chem. Phys.* **108**, 5855 (1998).

69. G. E. Walrafen, W.-H. Yang, and Y. C. Chu, *J. Phys. Chem. B* **103**, 1332 (1999).

70. D. M. Carey and G. M. Korenowski, *J. Chem. Phys.* **108**, 2669 (1998).

71. Y. Yasaka, M. Kubo, N. Matubayashi, and M. Nakahara, *Bull. Chem. Soc. Jpn.* **80**, 1764 (2007).

72. K. Yui, H. Uchida, K. Itatani, and S. Koda, *Chem. Phys. Lett.* **477**, 85 (2009).

73. Z. Wang, Y. Pang, and D. D. Dlott, *J. Phys. Chem. A* **111**, 3196 (2007).

74. Ph. Wernet, T. Testemale, J.-L. Hazemann, R. Argoud, P. Glatzel, L. G. M. Pettersson, A. Nilsson, and U. Bergmann, *J. Chem. Phys.* **123**, 154503 (2005).

75. J. Jonas, T. DeFries, and W. J. Lamb, *J. Chem. Phys.* **68**, 2988 (1978).

76. W. J. Lamb and J. Jonas, *J. Chem. Phys.* **74**, 913 (1981).

77. N. Matubayashi, Ch. Wakai, and M. Nakahara, *J. Chem. Phys.* **107**, 9133 (1997).

78. N. Matubayashi, Ch. Wakai, and M. Nakahara, *Phys. Rev. Lett.* **78**, 2573 (1997).

79. K. Yoshida, N. Matibayasi, and M. Nakahara, *J. Chem. Phys.* **127**, 174509 (2007).

80. K. Yoshida, N. Matibayasi, Y. Uosaki, and M. Nakahara, *J. Phys. Conf. Ser.* **215**, 012093 (2010).

81. N. Matubayashi, Ch. Wakai, and M. Nakahara, *J. Chem. Phys.* **340**, 129 (2001).

82. D. Sebastiani and M. Parrinello, *ChemPhysChem* **3**, 675 (2002).

83. T. Tsukahara, M. Harada, H. Tomiyasu, and Y. Ikeda, *J. Supercrit. Fluids* **26**, 73 (2003).

84. K. Okada, M. Yao, Y. Hiejima, H. Kohno, and Y. Kajihara, *J. Chem. Phys.* **110**, 3026 (1999).

85. M. Yao and Y. Hiejima, *J. Mol. Liquids* **96–97**, 307 (2002).

86. K. Okada, Y. Imashuku, and M. Yao, *J. Chem. Phys.* **107**, 9302 (1997).

87. R. B. Gupta, C. G. Panayiotou, I. C. Sanchez, and K. P. Johnston, *AIChE J.* **38**, 1243 (1992).

88. C. Panayiotou and I. C. Sanchez, *J. Phys. Chem.* **95**, 10090 (1991).

89. P. J. Smits, I. G. Economou, C. J. Peters, and J. de S. Arons, *J. Phys. Chem.* **98**, 12080 (1994).

90. I. G. Economou and M. D. Donohue, *Ind. Eng. Chem. Res.* **31**, 2388 (1992).

91. S. H. Huang and M. Radosz, *Ind. Eng. Chem. Res.* **29**, 2284–2294 (1990).

92. T. Vlachou, I. Prinos, J. H. Vera, and C. G. Panayiotou, *Ind. Eng. Chem. Res.* **41**, 1057 (2002).

93. Y. Marcus, *Phys. Chem. Chem. Phys.* **2**, 1465 (2000).

94. R. Lamanna, M. Delmelle, and S. Cannistrato, *Phys. Rev. E* **49**, 2841 (1994).

95. Y. Marcus, *J. Mol. Liquids* **81**, 101 (1999).

96. G. D. Ikomonou and M. D. Donohue, *AIChE J.* **32**, 1716 (1986).

97. A. G. Kalinichev, *Intl. J. Thermophys.* **7**, 887 (1986).

98. A. G. Kalinichev, and A. G. Kalinichev, *Ber. Bunsenges. Phys. Chem.* **97**, 872 (1995).

99. R. B. Gupta and K. P. Johnston, *Fluid Phase Equil.* **99**, 135 (1994).

100. L. B. Partay, P. Jedlovszky, I. Brovchenko, and A. Oleinikova, *J. Phys. Chem. B* **111**, 7603 (2007).

4

SCW AS A "GREEN" SOLVENT

The advent of so-called "green" solvents is a step in the amelioration of the ecological stress that modern industry has imposed upon human society. Solvents are necessary in chemical production and pharmaceutical and other formulations as well as in the disposal of hazardous chemicals, and until recent years organic solvents have been employed successfully for these purposes. Such solvents, however, are not generally "environmentally friendly," in that they may be toxic, flammable, and not entirely recyclable, so that their remnants could pollute water streams or subterranean water reservoirs. Several new types of solvents have been proposed for the replacement of the traditional organic solvents in processes. In an award address of the president of the American Chemical Society [1], concerning the Environmental Protection Agency (EPA) of the United States, he stressed the advantages accruing from the use of the "green" solvents. Such solvents include ionic liquids (that have very low vapor pressures), supercritical fluids (such as supercritical carbon dioxide and fluorocarbons) and water-based solvents. Each of such solvents has its advantages and drawbacks, in terms of remnant toxicity, flammability, vapor losses, cost, difficult operation conditions, corrosion problems, and so on, as well as limitations to the range of substances that can be dissolved and processed in it.

Supercritical water (SCW) is one of the "green" solvents that emerged in recent years and fulfils the requirements from such solvents to a high degree. SCW is certainly nontoxic and cannot pollute, as such, any water streams and reservoirs (except for heat pollution, if directly discharged). Neither is it flammable and if released to the atmosphere it has no deleterious effects.

Supercritical Water A Green Solvent: Properties and Uses, First Edition. Yizhak Marcus.
© 2012 John Wiley & Sons, Inc. Published 2012 by John Wiley & Sons, Inc.

As a material it is readily available and its cost is low. Recent examples of the application of SCW are a decentralized biomass refinery based on total oxidation by SCW [2] and SCW for environmental technologies [3].

The drawbacks of SCW as a solvent are related to the technical difficulties related to the high temperatures and pressures involved in its application, described in Chapters 1 and 2, and to the high demands from container and reactor material in order to avoid, or at least minimize, corrosion problems, discussed in Chapter 5. Some of the problems due to the extreme operation conditions required in the use of SCW are reduced when near-critical or hot liquid water under pressure is employed [4, 5], but the tunability feature of SCW, relating its density to the temperature and pressure, is lost in such media.

In the present chapter the solubilities of various substances in SCW are dealt with, so that the solvent properties of SCW are emphasized in view of its applicability to their processing. To be discussed are the solubility of substances that are gases at ambient conditions, of those—in general organic substances—that are liquid or solid, and of inorganic salts (including acids and bases). To round the discussion off, the properties of mixed solvents involving SCW are also described. A review on solubilities in supercritical fluids [6] did not deal with SCW, but presented a generally useful expression for correlating solubilities s in a fluid of density ρ and at a temperature T:

$$\ln s = k \ln \rho + a/T + b \qquad (4.1)$$

with three coefficients, the last, b, taking care of the units employed. In all the experimental studies reported in the following sections SCW must have been thoroughly deoxygenated, because oxygen readily oxidizes organic compounds in SCW, as is made use of in the supercritical water oxidation (SCWO) processes dealt with in Chapter 5.

4.1   SOLUTIONS OF GASES IN SCW

4.1.1   Phase Equilibria

The solubility of gases in water at ambient conditions is rather low, unless they react chemically, as do, for instance, HCl, NH_3, and CO_2. The solubility of the gas (subscript g) is described in terms of the Ostwald coefficient L_g, which is the ratio of the volume of a pure nonreactive gas dissolving at a given temperature and pressure in a given volume of pure water. The standard state molar volume of the gas for the standard pressure of $P^\circ = 0.1$ MPa ≈ 1 atm is $V_g^\circ = RT/P^\circ$, ignoring the very small term in the second virial coefficient of the gas (related to its self-association) at this pressure. The molar concentration

TABLE 4.1   The Solubilities of Some Gaseous Small Molecules in Water at 25°C at a Partial Pressure of the Gas of 1 atm, From Ref. [10] and the Minimal Critical Temperatures for Their Binary Mixtures with SCW, $t_{c\,min}$/°C

	$\log L$	$10^4\ (c_S/\mathrm{mol\cdot dm^{-3}})$	$10^4 x_S$	$t_{c\,min}$/°C
Helium	−2.024	3.86	0.0698	No $t_{c\,min}$
Neon	−1.958	4.50	0.0813	No $t_{c\,min}$
Argon	−1.468	13.95	0.252	364
Krypton	−1.216	24.85	0.449	357
Xenon	−0.978	42.95	0.776	344
Hydrogen	−1.718	7.82	0.1413	No $t_{c\,min}$
Oxygen	−1.507	12.72	0.2298	367
Nitrogen	−1.799	6.49	0.1173	366
Carbon monoxide	−1.632	9.54	0.1724	
Carbon dioxide	−0.082	338.2	6.111	266
Methane	−1.469	14.13	0.2507	353
Ethane	−1.344	18.84	0.3345	350
Propane	−1.436	14.97	0.2704	351
n-Butane	−1.527	12.16	0.2197	345
Tetrafluoromethane	−2.286	1.705	0.0382	

(in $1\,\mathrm{dm^3}$ of water) of a nonreacting gas is then $c_g = L_g P^\circ / RT = 12.03\,L_g /$ (T/K). Its mole fraction is $x_g = 1000\,L_g M_W P^\circ / \rho_W RT$, where $M_W / \rho_W = V_W$ is the molar volume of water, and at 25°C $x_g = 7.29 \times 10^{-4}\,L_g$.

Data bases of aqueous gas solubilities have been published, among others, by Crovetto et al. [7] over a very wide temperature range, $25 \leq t/°\mathrm{C} \leq 275$, and most recently by Battino and Clever [8]. Cabani et al. [9] were among those who related the solubility of gases in water to their properties, namely in terms of group contributions of the solute molecules. They reproduced the solubilities at 25°C of 209 hydrocarbons and monofunctional organic molecules in the gaseous state in terms of $\log L_g$(g in W) with a standard deviation of 0.09. The solubilities in water at 25°C of representative gases are shown in Table 4.1, with data taken from Wilhelm et al. [10]. The mole fractions of the dissolved gases are generally $x_g < 1 \times 10^{-4}$, except for the reactive gas, CO_2.

The solubilities of gases in SCW are, however, much larger; in fact, the small molecular gases are completely miscible with SCW at sufficiently high temperatures. Franck and coworkers presented the critical phase boundary curves of binary systems involving SCW and He [11], Ne [12], Ar [13, 14], Kr [12], Xe [15], H_2 [16], N_2 [17], O_2 [18], CO_2 [19, 20], CH_4 [21, 22], and C_2H_6 [23]. Above the critical temperature surfaces $T_c(P, x_g)$ for these binary mixtures the two components are miscible.

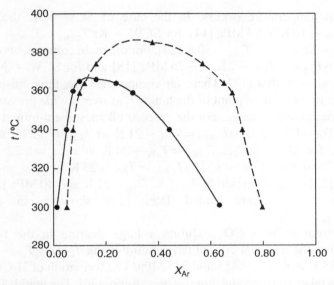

FIGURE 4.1 Binodal curves for argon in SCW at two pressures: —●— at 50 MPa and
–▲– at 250 MPa, drawn from data in Fig. 6 in Ref. [24].

The solubilities of the gases in SCW are described in the three-dimensional $T - P - x$ representation by the critical bimodal surfaces $T_c - P_c - x_c$. An example of isobaric projections of the surface for Ar in SCW is shown in Fig. 4.1. Above the surface the system consists of a single fluid phase, whereas below it there are two fluid phases (liquid and gas). As the pressure is increased the surface is seen to widen, so that the miscibility at a given temperature diminishes. The upper consolute temperature also moves to higher argon contents.

The systems SCW + He, SCW + Ne, and SCW + H_2 exhibit gas–gas immiscibility of the first kind, Type IIIa of the van Konynenburg and Scott classification [25]. According to the Temkin theory [26], since they obey the relations $V_{cg} \geq 0.42V_{cw}$ and $T_{cg}V_{cg} < 0.052T_{cw}V_{cw}$, they do not exhibit a minimum in the temperature–pressure critical curve $T_c(P)$ at any x_g. The critical curve for the coexistence for two phases rises with the gas contents, in the case of He as $T_c/K = 647 + 19.0\,x_{He} + 322\,x_{He}^2$ and $P_c/MPa = 22 + 32\,x_{He} + 1143x_{He}^2$ up to $x_{He} = 0.35$, where $T_c = T_{cw} + 49$ K [11] ($T_{cw} = 647$ K). In the case of SCW + $H_2 T_c$ remains close to that of water, rising only to $T_{cw} + 7$ K at $x_{H_2} = 0.38$ [16]. Scalise [27] confirmed the steep rise of pressure with temperature of the critical curve for SCW + He and SCW + H_2 without showing a minimum.

For the other gases a phase behavior according to Type IIIb of the van Konynenburg and Scott classification [25] takes place, in that a minimum in the critical temperature occurs. In the case of SCW + Ar the minimal $T_{c\,min} = T_{cw} - 10$ K at 84 MPa [14], for SCW + Kr $T_{c\,min} = T_{cw} - 17$ K, and for SCW + Xe $T_{c\,min} = T_{cw} - 30$ K [24]. For the main components of air: for SCW + O_2 $T_{c\,min} = T_{cw} - 7$ K at $\sim$75 MPa [18] and for SCW + N_2 $T_{c\,min} = T_{cw} - 8$ K at $\sim$75 MPa [17]. These air components are, thus, miscible with SCW close to the critical point of the latter, T_{cw}, over a wide pressure, hence density (see Table 2.1) range. For the gaseous alkanes, the minimal temperatures are for SCW + CH_4 $T_{c\,min} = T_{cw} - 21$ K at $x_{CH_4} = 0.245$ and $\sim$100 MPa [22], for SCW + C_2H_6 $T_{c\,min} = T_{cw} - 24$ K at $x_{C_2H_6} = 0.22$ and $\sim$73 MPa [23], and for SCW + C_4H_{10} $T_{c\,min} = T_{cw} - 23$ K at $x_{C_2H_6} = 0.11$ and $\sim$38 MPa [23] or later revised to $T_{c\,min} = T_{cw} - 22$ K and 40 MPa [28]. Data for SCW + C_3H_8 were added later [29] showing in a figure $T_{c\,min} = T_{cw} - 29$ K.

The system SCW + CO_2 exhibits a huge decline in the two-phase critical point, the minimal critical temperature being $T_{c\,min} = T_{cw} - 108$ K = $T_{CO_2} + 235$ K at $x_{CO_2} = 0.415$ and 245 MPa. The formation of H_2CO_3 in this system was taken into account, but it is quite minor [20]. The nonideality of the single phase mixtures was presented by Franck and Tödheide [19] in terms of the compressibility factor $Z = PV/RT$ as a function of the composition. Greenwood [30] presented activity (fugacity) data of the components at 450–800°C. According to calculations of Kerrick and Jacobs [31] at 400°C and a pressure of 2 GPa unmixing takes place at $0.20 \leq x_{CO_2} \leq 0.45$ and at a pressure of 3 GPa at $0.2 \leq x_{CO_2} \leq 0.6$. However, at 500°C the deviations from ideality are smaller, so that they do not lead to unmixing at even 3 GPa, the deviations decreasing further as the temperature is increased to 800°C. Duan et al. [32] showed unmixing that occurs for SCW + CO_2 at 300°C (still above $T_{c\,min} = 266°C$) at $0.15 \leq x_{CO_2} \leq 0.45$ and 50 MPa, as arising from their equation of state, turning to near ideality at 700°C.

A refrigerant mixture of ammonia and water, $x_{NH_3} = 0.2607$, was studied at supercritical states by Polikhrinidi et al. [33] to yield PVT and critical data at pressures up to 28 MPa.

The minimal critical temperatures for the binary mixtures of gases in SCW are shown in Table 4.1.

4.1.2 Interactions in the Solutions

Once a gaseous solute is dissolved in SCW it interacts with the water molecules according to its inherent properties. For instance, argon interacts mainly by dispersion forces, although the dipoles of the water molecules may polarize the argon atoms to some extent. Cummings et al. [34, 35] applied computer simulations with the SPC water model to obtain the pair correlation functions $g(Ar-W, r)$, The excess of water molecules in a correlation region near an argon atom extending up to a distance R from its center is

$$N(R) = \rho_{W_0} \int^R 4\pi r^2 (g(\text{Ar}-\text{W}, r) - 1) dr \qquad (4.2)$$

The net effect, $N(R) < 0$, is that argon is a repulsive solute in SCW at the reduced temperatures $T_r = 1.05$ and 1.00 and reduced densities of $\rho_r = 1.00$ and 1.50, respectively. A deficit of nearly three water molecules was noted around argon relative to their mean number around a water molecule. Similar conclusions were obtained from an X-ray absorption fine structure (XAFS) study of 1.5 mol% krypton in SCW [36], although the repulsive nature of this noble gas atom is somewhat less than that of argon, being more polarizable, thus more attractive.

The Raman spectrum of hydrogen in SCW shows line broadening of both the Q-branch and the S-branch, being most noticeable at low J values, and due to inelastic collisions and reorientation of the H_2 molecule, respectively [37]. At liquid-like densities of SCW the pure collisional mechanism for the line broadening breaks down and an effect of the hydrogen bonded structure of the SCW is noted.

Bermejo et al. [38] applied the multiparameter Anderko–Pitzer equation of state (see Section 2.1.3) to solutions of the components of air, that is, oxygen and nitrogen, in SCW to obtain thermodynamic quantities pertaining to such solutions. Binary interaction parameters between water and oxygen and water and nitrogen were fitted with vapor–liquid data at supercritical temperatures. The modeling was applied to mixtures of 65 mol% water and 35 mol% air and were calculated for SCW conditions at 800°C for the densities and for several temperatures up to 800°C for the heat capacities. Seminario et al. [39] deduced from the van der Waals a and b parameters for water and oxygen that this pair of molecules would be very weakly attractive in SCW at 577°C and 25 MPa (at a gas-like density of 74 kg·m^{-3}). Their molecular dynamics (MD) simulations shows that under these conditions oxygen reduces the clustering of the water molecules. Further MD simulations by Omori and Kimura [40] also showed expansion of the water by oxygen near the critical point, leading to depletion of water molecules near the oxygen ones. The Raman spectrum of oxygen in SCW was measured by Sugimoto et al. [41] at 380–500°C and gas-like densities of 0.9–31 kg·m^{-3}. The rotational relaxation of the O_2 molecules in such SCW does not differ much from that in gaseous O_2, but at densities above 15 kg·m^{-3} and at the lower temperatures depletion of water densities is found from the Raman data, in agreement with the MD simulations. A similar conclusion was reached by Yui et al. [42] from examination of the Raman O–H stretching frequencies in 1 mol% O_2 in SCW near the critical point.

A diatomic virtual probe solute was used for MD simulations of its solvation in SCW by Duan et al. [43]. The solvation structures varied drastically with the dipole moment of the solute, especially when the density of the SCW was low. The solvent relaxation at both high and low densities was explored.

Computer simulation calculations [44] with SCW + CO_2 at $x_{CO_2} = 0.0092$ and 0.092, 400°C, and 130 MPa (densities of 0.463 and 559 kg·m^{-3}) showed that percolation does not take place (see Sections 3.2 and 3.4), contrary to pure SCW at the same temperature and at a similar density (580 kg·m^{-3}). There are only $\chi \sim 1.2$ hydrogen bonds per water molecules in the presence of CO_2, which is below the percolation (continuous hydrogen bonded network) limit. This lack of percolation is ascribed to the large excess volume caused by the presence of the solute. The Raman spectrum of 3–20 mol% CO_2 in SCW, and in particular the hot bands of the symmetric stretch vibration have been measured by Brown and Steeper [45] at 390–540°C and 34.5 MPa. The integrated intensity ratio of the hot and fundamental bands at 1409 and 1388 cm^{-1} was found to be linear with the temperature. It could serve as a thermometric indicator in SCW with an accuracy of ±6%, not being affected by the density or by the presence of water, oxygen, or hydrocarbon molecules.

Solutions of methane, as an archetypical hydrophobic solute, were studied in SCW by Matubayashi and Nakahara [46] and by Hernandez-Cobos and Vega [47] by means of computer simulations. The former authors used the SPC/E model and nominally 400°C ($T_r = 1.09$) and densities of 200, 400, 600, 800, and 1000 kg·m^{-3} (pressures from 22.5 to 230.3 MPa) and the latter used the TIP4P model and nominally 600 K ($T_r = 1.02$) and densities of 250, 500, and 997 kg·m^{-3}. Both sets of authors concluded that the enthalpy of solution is relatively small and positive and counteracts the dissolution, but the entropy is large, negative, and becomes more negative as the pressure and density are increased. Matubayashi and Nakahara [46] applied the scaled particle theory for the estimation of the effective diameter of a water molecule, finding it to decrease from 0.276 nm at the highest density to only 0.172 nm at the lowest (much less than the van der Waals diameter), the reduction being related to the spatial inhomogeneity of the SCW.

4.2 SOLUTIONS OF ORGANIC SUBSTANCES IN SCW

4.2.1 Phase Equilibria

It is often stated in the literature that an important feature of SCW is that the solubilities of organic compounds in it, whether polar or nonpolar, is very high or even that some of them are miscible with SCW at all proportions. However, the availability of concrete data to substantiate such claims is very meager.

Schneider et al. measured the phase equilibria of mainly aromatic hydrocarbons and some of their derivatives in SCW [48–52]. The bimodal curves are rather broad in the $P_c - x_c$ representations at the minimal critical temperature, $T_{c\,min}$, so that it is difficult to specify a composition at which this minimum takes place. Also, the compositions in the earlier papers are given in mass%, but in the ones in the late 1970s this was not specified. For SCW + benzene the values of $T_{c\,min}/K$ range from 292.8 to 294.8 at 30.5–54.6 mass% [48]. This value was fixed at 294.2 in Ref. [49], where the values of $\Delta T_c = T_{cW} - T_{c\,mixture}$ were reported to diminish gradually from 80 K for benzene, to 61 K for toluene, 57 K for o-xylene, 55 for ethylbenzene, and 47 K for 1-propylbenzene and 1,3,5-trimethylbenzene. The critical pressures at the minimal temperature range from 16 to 25 MPa. For benzene in supercritical D_2O, $T_{c\,min}$ is higher than for the light water system by 4 K [50]. The values of ΔT_c for the SCW + cyclohexane and SCW + n-heptane are 28 K at 30 mass% and 35.5 MPa and 23 K at 33 mass% and 25 MPa, respectively, and for SCW + biphenyl the corresponding values are 55 K, 29 mass%, and 12 MPa [50]. For tetralin and cis- and $trans$-decalin the critical parameters are very similar, $\Delta T_c = 25$ K [51] but for fluorobenzene and p-difluorobenzene on SCW ΔT_c is again much larger, the same as for benzene itself, 80 K.

The study by Yling et al. [28] of mixtures of SCW with n-butane and with n-hexane continued the series of mixtures of SCW with aliphatic hydrocarbons that are gaseous at room temperature by Franck et al., dealt with in Section 4.1.1. In the case of hexane, the minimal critical temperature is at $T_{cW} - 19$ with $x_{C_6H_{14}} = 0.06$ and 30.5 MPa, in agreement with de Loos et al. [53] who reported similar data. Above the critical curve SCW and n-hexane are miscible.

A comprehensive and systematic study of the critical phase equilibria of SCW, in terms of $T(P)$ curves, with n-alkanes (C_nH_{2n+2} with all n from 1 to 12 and the even n from 14 to 36 except 34) was reported by Brunner [54]. He confirmed the above data for n-hexane, and added $T_{cW} - 22$ at 32.5 MPa for n-pentane. The values of $\Delta T_c = T_{cW} - T_{c\,mixture}$ diminish gradually from 25.4 for $n = 3$ to 6.5 for $n = 26$ ($\Delta T_c = 36.2 - 1.837n + 0.0364n^2$) and then abruptly for higher n: 3.0 at $n = 28$, -5.9 at $n = 30$, -12.9 at $n = 32$, and -17.1 at $n = 36$. In the latter three cases the critical point for the mixtures is higher than that of water. The critical minimum temperature appears to approach more and more toward the water-rich end of the phase diagram, as n increases, but no phase compositions were determined in this study.

Brunner et al. [55] reported a similar study of mixtures of SCW with many aromatic hydrocarbons. Again, only $T(P)$ curves but no phase compositions were determined in this study, so that actual solubilities are unavailable.

The results essentially confirm the earlier data of Schneider et al. presented above. Methyl benzenes exhibit a large diminution of the minimal critical temperature: $\Delta T_c = 78$ K for benzene, 63 K for toluene, 56 K for o-xylene, 52 K for p-xylene, and 45 K for 1,3,5-trimethylbenzene. Larger aromatic hydrocarbons have smaller values of ΔT_c: 28 K for naphthalene, 23 K for 1-methylnaphthalene, biphenyl, diphenylmethane, and diphenylethane, and only 7 K for 1,4-diphenylbenzene (triphenyl). Also studied were acenaphthene, fluorene, phenenthrene, anthracene, pyrene, tetralin, and indene among the aromatic hydrocarbons. Further included in this study were derivatives such as phenol, benzophenone, diphenylether, 1-chloronaphthalene, and benzyl, as well as the nonaromatic cyclopentane and *trans*-decalin. All these have ΔT_c values in the range of 8–31 K and exhibit Type-II critical curves of the van Konynenburg and Scott classification [25], the liquid–gas critical curve being above the liquid–liquid one. Bröllos et al. [50] previously studied the SCW + cyclohexane system, where $\Delta T_c = 28$ K was found, SCW + n-heptane and SCW + biphenyl showing similar behavior.

The cross virial coefficients of SCW with methane, n-hexane, n-octane, and benzene were determined by Abdulagatov et al. [56] from their gas phase densities up to 400°C and 100 MPa. Abdulagatov et al. later added measurements of the $PVTx$ properties of mixtures of SCW + n-heptane [57] and of SCW + n-pentane [58].

As an illustration, the critical temperature $T_c(x)$ curve for gas–gas equilibria in the water + n-hexane system are shown in Fig. 4.2. The narrow

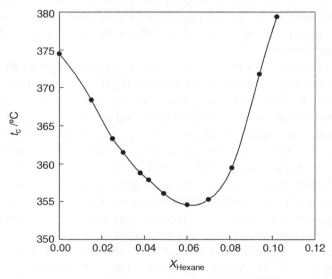

FIGURE 4.2   Binodal $T_c(x)$ curve for n-hexane in SCW drawn from data in Ref. [53].

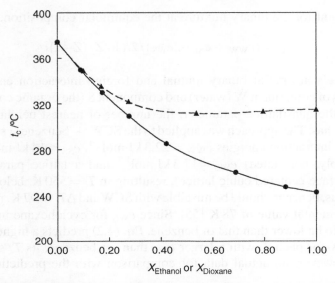

FIGURE 4.3 Critical temperature curve $T_c(x)$ for the SCW + ethanol system –•– and the SCW + dioxane system –▲– from data in Ref. [59].

region near the SCW side, where a single phase occurs, should be noted. This is typical of other SCW + hydrocarbon systems too.

Many hydrophilic polar solutes appear to be miscible with SCW at all proportions, a single fluid phase existing for them above the critical temperature curve $T_c(x)$ extending between the critical temperatures of the two pure components. This was established with mixtures of SCW with methanol, ethanol, 1,4-dioxane, tetrahydrofuran, and acetone [59]. This is illustrated in Fig. 4.3 for the SCW + ethanol and SCW + dioxane systems. Note that in the latter system there is a shallow minimum in the curve, occurring also in the SCW + tetrahydrofuran system. In the cases of SCW + methanol and SCW + ethanol the $T_c(x)$ curves were substantially confirmed by other workers, as is shown in the work of Abdulagatov and coworkers [60, 61]. For SCW + ethanol a Redlich–Kister type expression is valid [60]:

$$T_c(x) = (1-x)T_{cW} + xT_{cEtOH} + x(1-x)[129.8 - 49.9(1-2x) + 4.83(1-2x)^2]$$
$$(4.3)$$

Similar expressions were given for the critical pressures and densities.

Biswas and Bagchi [62] attempted to explain the high solubility of organic compounds in SCW in terms of a quasilattice quasichemical model [63]. It predicts a symmetrical critical curve (with respect to composition) and upper critical point for the binary mixture at the equimolar composition:

$$T_c = (e_{WW} + e_{SS} - 2e_{WS})/2R \ln[Z/(Z-2)]$$ (4.4)

Here the e's are molar binary mutual and foreign interaction energies of molecules of component W (water) and component S (the organic compound) and Z is the quasilattice parameter, the number of nearest neighbors each lattice site has. The approach was applied to the SCW + benzene system, with the molar interaction energies $e_{WW} = 22.3\,kJ\cdot mol^{-1}$, $e_{SS} = 14\,kJ\cdot mol^{-1}$, and (presumably, not stated) $e_{WS} = 17.3\,kJ\cdot mol^{-1}$ and a lattice parameter of $Z = 12$ (a face-centered cubic lattice), resulting in $T_c = 550\,K$, below that of water. Thus, benzene should be miscible with SCW and $\Delta T_c = 97\,K$, more than the experimental value of 78 K [55]. Since e_{WS} for cyclohexane in water is expected to be lower than that of benzene, Eq. (4.2) predicts a higher T_c, and conversely for phenol, with a larger e_{WS} than for benzene, its T_c should be lower. However, no actual data for comparison with the predictions were presented.

The solubility expression for the solubility according to the Hildebrand solubility parameter concept is

$$\ln x_{S\,sat} = -V_S \varphi_W^2 (RT)^{-1} (\delta_W - \delta_S)^2$$ (4.5)

where V_S is the molar volume of the solute, φ_W is the volume fraction of SCW in the saturated solution, and the δ's are the solubility parameters. If the solubility parameters of potential solutes, such as condensed-ring or poly-chlorinated aromatic hydrocarbons that are of environmental importance, are assumed to be not very sensitive to the temperature, then the solubility parameter of water may be used in order to estimate the solubility of the solute. Marcus [64] reported the value of $\delta_W = 13.7\,MPa^{1/2}$ at $T = 1.2T_r$ and $\rho = 1.0\rho_r$. He showed that δ_W is

$$\delta_W/MPa^{1/2} = 15.1[1 + 0.937(1 - T_r^{1/2})]T_r^{1/4}\rho_r$$ (4.6)

that is, it is proportional to ρ_r and its temperature dependence is rather mild. Again, no comparisons of predicted and experimental solubilities were presented.

Artemenko and Mazur [65] dealt with the phase diagrams of mixtures of environmentally important organic compounds (polycyclic aromatic hydrocarbons (PAHs), polychlorinated biphenyls (PCBs), and polychlorinated dibenzodioxins and furans) with SCW. They based their approach on global phase equilibrium diagrams [25] and cubic equation of states (EoS), in particular the Redlich–Kwong–Suave (RKS) one (see Section 2.1.2). The a, b, and α parameters of this EoS for SCW at 400°C and pressures from 15 to 45 MPa were presented. For binary mixtures the a_{ij} interaction parameters involve those of the pure components modified by the factor $(1 - k_{ij})$, which, in turn, are obtained from correlations with the known octanol/water partition coefficients K^O_W. Phase envelops (P, T) are presented for SCW + pyrene, biphenyl, and the pesticides DDT and 2,4-D.

4.2.2 Interactions in the Solutions

Solvatochromic probes have found extensive use for the estimation of properties, such as solvation ability, polarity, and hydrogen bonding ability, of solvents at ambient conditions. Some such use has also been made of suitable probes in SCW. It is difficult, however, to generalize results from a specific solvatochromic probe in SCW, which probes the properties of its local environment and induces clustering of water molecules around it, to the overall properties of the SCW, due to the selective solvation phenomenon. There exists an additional problem with solvatochromic probes in SCW, in that they tend to be unstable and if air is not carefully excluded they are oxidized, with oxidation products that have similar but not exactly the same solvatochromic properties.

The paper by Bennet and Johnston [66] appears to be the first in which UV-visible spectroscopy was applied to probes in SCW. Since nitroaromatics were found to be unstable in SCW, the authors used benzophenone and acetone as their π^* (polarity/polarizability) probes in SCW up to 440°C. Acetone is miscible with water as a supercritical fluid mixture above its critical temperature of 238°C [59], so that a wide range of temperatures is available for its spectroscopic study. The $n–\pi^*$ band of acetone is red-shifted from its position, 262 nm, in ambient water or pressurized water to 280 nm at 380°C. Above 28 MPa the band position is independent of the pressure but is affected by both solvatochromism and thermochromism. The latter effect is removed by spectral data of acetone vapor in argon gas, and the thermochronic-corrected band maximum spectral shift in SCW becomes:

$$\Delta v(T, \rho) = v(T, \rho) - [42.8 - 1.71(t/°C)] \tag{4.7}$$

where v is in wave numbers and the second term is the thermochronic-correction. Independent of the temperature, as the reduced density ρ_r of SCW diminishes from 3.2 to 2.0 there is an appreciable red-shift of the band maximum, ascribed to decreased hydrogen bonding donor strength of the water, but even if the density is decreased to $\rho_r = 0.5$ about one half of $\Delta v(T, \rho)$ is due to the hydrogen bonding, the other half to physical effects. For $0.5 \le \rho_r \le 1.5$ a plateau in the $\Delta v(T, \rho_r)$ curve is observed. It is concluded that the local density of water molecules near the acetone ones is much larger that the bulk density and changes little when the latter diminishes.

For benzophenone in SCW, an increase in temperature causes a decrease in the solvent density relative to ambient water and reduces its polarizability per unit volume, causing a blue shift of the absorption spectrum of the benzophenone, as is known for conventional solvents. Conversely, an increase in pressure (and density) in SCW causes a red shift, by increasing the polarizability density. The probe does, however, affect the density of the nearby SCW, augmenting it by interactions of the highly polarizable probe with the dipoles of the water and by hydrogen bond donation from the water to the carbonyl group. As the reduced density of the SCW at 380°C is diminished from $\rho_r = 1.5$ the spectral shift corresponds to an increase in the local density, up to $\rho_r^{local} = 1.3$ at $\rho_r = 1.0$. The spectral shift in SCW at 400°C and 34.5 MPa was reproduced by Monte Carlo computer simulation with a polarized benzophenone model [67].

The acid–base reaction of dilute solutions of β-naphthol in SCW at 400°C was studied by Xiang and Johnston [68] by UV-spectroscopy. The acidic form has a maximum value at 326.7 nm whereas the basic form (naphtholate anion) has shifted to 370 nm. The proton exchange between naphtholate (A^-) and hydroxide anions: $HA + OH^- \leftrightarrows A^- + H_2O$ takes place with no net change in the charges in the system, hence the equilibrium constant K_{BHA} is not sensitive to the changes in the permittivity of the system as the density of the SCW is varied, but only to the interactions with the solvent. At 400°C log K_{BHA} decreases from 3.1 at 200 kg·m^{-3} to 1.2 at 1000 kg·m^{-3}. The negative charge is preferred on the larger naphtholate anion than the smaller hydroxide one and the reaction is endothermic. The equilibrium constant for the ionization of β-naphthol, $HA \leftrightarrows A^- + H^+$, is $K_a = K_{BHA}K_W$, where K_W is the ion product of water, dealt with in Section 2.5, and the ionization reaction is exothermic.

Lu et al. studied the application of the Kamlet–Taft solvatochromic parameters to probes in near critical water, along the saturation curve up to 275°C, [69, 70] but not to SCW. The polarity/polarizability parameter π^*, measured with 4-nitroanisole, diminishes nonlinearly (concave downward) from its value of 1.09 at ambient conditions to 0.69 at 275°C, in direct relation to the decreasing density of the water (from 997 to 663 kg m^{-3}). The hydrogen bond donation ability α, measured with a dichloro-substituted pyridinium-N-phenoxide probe, diminishes nonlinearly (concave upward) in this range from 1.16 to 0.84. The hydrogen bond acceptance ability β was measured with 4-nitroaniline and N,N-dimethyl-4-nitroaniline increases slightly in this range, but these probes are unstable and eventually hydrolyze.

Oka and Kajimoto [71], however, did study the spectral properties of 4-nitroaniline, N,N-dimethyl-4-nitroaniline, and 4-nitroanisole in SCW at 380, 390, and 410°C. It was necessary to employ a flow system, so as to restrict the residence time of the indicators in SCW, due to their instability at SCW conditions. 4-Nitroaniline has an absorption peak at 380 nm in ambient water that is blue-shifted in SCW at 380°C, the more the lower the density of the SCW (to ~300 nm at a density of 123 kg·m^{-3}). Similar effects were obtained with the other indicators and at the other temperatures, but the blue shift is much less at densities above ~300 kg·m^{-3}. It was concluded from a comparison of results from the three indicators that no specific hydrogen bonding between the water molecules of SCW with either the nitro-group or the amino group of 4-nitroaniline takes place, and that the spectral shifts are due to the extent of hydrogen bonding in the SCW itself. It was also found that the wavenumber of the absorption peak is related to the bulk permittivity of the SCW at 410°C as

$$\nu_{max}/cm^{-1} = 12740(\varepsilon-1)/(2\varepsilon+1) \qquad (4.8)$$

However, at 380 and 390°C in SCW there are deviations downward from the linearity of ν_{max} with $(\varepsilon-1)/(2\varepsilon+1)$. This is ascribed to local enhancement of the permittivity near the solute by clustering of solvent molecules around it. Thermal movements at the higher temperature (410°C) cause the clusters to dissipate, hence the observed linearity of Eq. (4.8). The density fluctuations in supercritical fluids, responsible for the clustering, are indeed found to diminish considerably as the temperature is raised and moved away from the critical point. Contrary to the conclusion of Oka and Kajimoto [71] concerning the absence of direct hydrogen bonds between the water molecules and the amino group of p-nitroaniline, a recent Raman spectroscopic study of Fujisawa et al. [72] did find such hydrogen bonds to be formed in SCW. The decay rate of an excited 4-nitroaniline molecule in SCW at 391°C and 40.1 MPa was measured by Osawa et al. [73]. The decay of the hot band was more moderate than predicted by collision theory, a fact that suggested the effects of the local density enhancement around the solute.

The UV-visible spectrum of quinoline in SCW (380–430°C, up to 40 MPa) was measured by Osada et al. [74]. They concluded that at conditions where the compressibility of water is large ($0.5 < \rho_r < 1.5$) the local density of water around the quinoline is lower than in bulk water, that is, *negative* solvation takes place. The hydrogen bonding between water and quinoline is strongly diminished. On the contrary, local density augmentation occurs in SCW around the exciplex between acetophenone and tetramethylbenzidine at 380–410°C [75, 76], the more the augmentation, the lower the density. In fact, at a bulk density of 150 kg·m^{-3} ($\rho_r \sim 0.65$) the local density of the water near the exciplex is raised to 270 kg·m^{-3}. Similarly, the local water density is augmented in solutions of pyridazine in SCW as found by Minami et al. [77] from UV-visible spectroscopy at 380–420°C. At 380°C and a bulk density of 200 kg·m^{-3}, there are 30% as many hydrogen bonds

($\langle n_{HB} \rangle = 0.52$) as in ambient water, considerably more than in the absence of the solute (Table 3.3).

Minami et al. [78] measured the UV spectrum of 4-nitroanisole in SCW, again in a flow system in order to minimize the decomposition, and calculated the Kamlet–Taft solvatochromic parameter π^*, defined as

$$\pi^* = (v_{max}/cm^{-1} - 33,985)/(31,612 - 33,985) \qquad (4.9)$$

where 33,985 is v_{max}/cm^{-1} in cyclohexane ($\pi \equiv 0$) and 31,612 is v_{max}/cm^{-1} in dimethylsulfoxide ($\pi^* \equiv 1$) at ambient conditions. A linear dependence of π^* on the density was found for SCW at 400 and 420°C:

$$\pi^* = -(0.71 \pm 0.06) + (1.77 \pm 0.09)\rho \qquad (4.10)$$

but at 380 and 390°C deviations from linearity (curves concave downward) were observed, in agreement with the findings of Lu et al. [70] for saturated and near-critical water reported above. Local density augmentation (clustering) at 380 and 390°C was invoked to explain these deviations, an effect no longer present at ≥ 400°C. The density augmentation parallels the isothermal compressibility of SCW at 380 and 390°C, having a maximum at the same bulk density, $300\,kg \cdot m^{-3}$. The main interaction between the indicator molecules and the water are dipole–dipole ones rather than hydrogen bonding, and these are weaker than the hydrogen bonding of the water molecules among themselves in SCW. The local density augmentation near N,N-dimethyl-4-nitroaniline at 380°C was found to be somewhat larger than around 4-nitroanisole, but was the same for N,N-dimethylaminobenzonitrile as that of the latter. Osawa et al. [79] used Raman spectroscopy of the $C \equiv N$ stretching vibration to study p-aminobenzonitrile in SCW and supercritical methanol, finding a red shift with increasing solvent density. However, at $\rho_r > 2$ there occurred a blue shift. These UV-visible and Raman spectroscopic methods have been used to study indicator probe molecules in SCW, and in particular the local density enhancement near the critical temperature.

Nuclear magnetic resonance, specifically the ^{13}C-NMR chemical shift, $\delta(^{13}C=O)$, in acetone dissolved in SCW up to 400°C was used for the investigation of the hydration and hydrogen bonding in this system, supplemented by Monte Carlo computer simulation by Takebayashi et al. [80]. The NMR chemical shifts, both $\delta(^{13}C=O)$ and $\delta(^1H_2O)$ decrease somewhat as the temperature increases isochorically in SCW, but more appreciably as the density diminishes isothermally from liquid-like densities ($600\,kg \cdot m^{-3}$) to gas-like ones ($100\,kg \cdot m^{-3}$). The nonlinear density dependence of $\delta(^{13}C=O)$ from 196.7 ppm for an isolated acetone molecule to 208.2 at a density of

$600 \, \text{kg·m}^{-3}$ (a concave downward curve) parallels closely the blue shift of the UV absorption peak reported by Bennett and Johnston [66]. The results are interpreted in terms of competition between hydrogen bonding of water molecules to acetone (up to 0.7 at the highest density) and among themselves, in a manner confirmed by the computer simulations. Fonseca et al. [81] applied Monte Carlo computer simulations and quantum mechanical theoretical calculations to study the chemical shifts of acetone, $\delta(^{13}C=O)$ and $\delta(C=^{17}O)$, in SCW at 400°C and 34.5 MPa with the SPC/E model for water and included solute polarization in the computations.

Nieto-Draghi et al. [82] studied the interactions between benzene and SCW at 400°C at densities of 660 and $995 \, \text{kg·m}^{-3}$ and two mole fractions of benzene: 0.10 and 0.21 by means of molecular dynamics simulations with the SPC/E model of water and an anisotropic united atom model for benzene. It was concluded that at a density of $660 \, \text{kg·m}^{-3}$ almost half of the benzene molecules have one hydrogen bond with a water molecule. This hydrogen bond is longer lived than the mean lifetime of hydrogen bonds between water molecules in SCW.

The conditions for good miscibility of SCW with heavy oils were briefly studied by Morimoto et al. [83] in terms of the relative permittivity that should be in the range $2.2 \leq \varepsilon_r \leq 10.4$ and the Hansen solubility parameter (Section 2.6) for hydrogen bonding $\delta_h < 10.4 \, \text{MPa}^{1/2}$.

4.3   SOLUTIONS OF SALTS AND IONS IN SCW

4.3.1   Solubilities of Salts and Electrolytes

The solubilities of gases and organic solutes in supercritical water is manifold larger than their solubilities in water at ambient conditions. This is ascribed by some authors to the drastic decrease in the permittivity, hence of the polarity, of SCW compared with ambient water. However, the diminished permittivity is only a consequence of the breakdown of the hydrogen-bonded network. This network demands the expenditure of considerable energy to create a cavity to accommodate a solute that interacts only weakly with the water molecules by van der Waals or dipolar interactions.

Contrarily, the solubilities of ordinary salts are greatly diminished in SCW at gas-like densities ($\leq 200 \, \text{kg·m}^{-3}$) compared to their solubilities in water at ambient conditions. The large lattice energies (enthalpies) of crystalline salts must be compensated by large hydration energies (enthalpies) of the ions in order for solubility to take place, the entropies playing only a minor role for common salts. In this case the low permittivity does play a more direct role, in that ion association is enhanced as the permittivity is decreased, and the ion pairs so formed are less well hydrated than the ions. The lower density of SCW, unless the pressure is very high, also plays a role, as fewer water molecules are available for the hydration of the ions to overcome the lattice energies of crystalline salts. The entropy of solution then becomes of greater importance in determining the solubility.

As the density of the SCW is increased, however, the solubility of many salts increases too, with hydration and ion pairing competing, but both work against the lattice energy that must be invested to effect solubility. In some cases complete miscibility is attained, in other cases a two-fluid equilibrium occurs above the melting point of the salt. Mineral acids, such as HCl, HNO_3, H_2SO_4, and H_3PO_4, appear to be miscible with SCW at all proportions, since no phase diagrams indicating limited miscibility seem to have been reported.

Valyashko [84] reviewed the phase equilibria of water–salt systems at high temperatures and pressures. Many of the systems mentioned pertain to SCW conditions and appreciable salt concentrations, and the original publications should be consulted for detailed phase diagrams. Valyashko and Urusova [85] presented a small scale figure with the solubilities $x_{salt}(t)$ of Na_2CO_3, M_2SO_4 (M = Li, Na, K), and $BaCl_2$ in SCW ranging up to ca. 15 mol% (classified as type 2) and of NaCl, NaBr, KCl, and $Sr(NO_3)_2$ ranging up to at least 30 mol% (type 1), increasing with the temperature. The pressures involved in the saturated solutions are lower than those in SCW at corresponding temperatures but can be as large as 40 MPa for type 1 salts, having a maximum at some temperature, again shown in a small scale figure. The phase diagrams of such systems can be complicated by liquid immiscibility and other features.

The solubilities of salts in SCW have been studied to a large extent in connection with the SCWO process (see Chapter 5), in which organic substances containing atoms of halogens, sulfur, nitrogen, and phosphorus are converted eventually to halide, sulfate, nitrate, and phosphate salts. Since sodium hydroxide is generally employed as a solute in the SCWO process, these are then sodium salts of these anions. Few other salts have been studied with respect to their solubilities in SCW. Of the sodium salts mentioned, the chloride has received the major attention.

Crystalline sodium hydroxide (β-form) melts at 319.1°C, and above this temperature it is completely miscible with water [86, 87]. The solid–liquid phase diagram below this temperature can be modeled with the BET method [88], requiring the latent heat of fusion of β-NaOH (312 J mol^{-1}). The liquidus curve at saturation pressure, up to $x_W = 0.42$, is

$$t/°C = 319 - 44x_W - 1204x_W{}^2 \tag{4.11}$$

above which there is a single liquid phase. The monohydrate melts at 62°C and is at equilibrium with the liquid mixture at the equimolar composition.

At higher pressures the single fluid phase extends to SCW. For instance, at 100 MPa and 400°C the conductivity of NaOH in SCW [87] and its molar volume [89] were measured over the entire composition range.

The solubility of sodium chloride in SCW has attracted the major attention of investigators over many years. Bischoff and Pitzer [90] and Bischoff [91] compiled and critically evaluated the phase equilibria of NaCl in SCW up to 500°C and Anderko and Pitzer [92] referred to further studies at higher temperatures. A cubic expression fits the solubility of NaCl in SCW up to the melting point of the salt:

$$x_{\text{NaCl satd.}} = 0.090 + 1.1183 \times 10^{-7}(t/°C)^2 + 1.6643 \times 10^{-9}(t/°C)^3$$

$$(4.12)$$

The pressure of the saturated solution increases from $\sim$12 MPa at 380°C ($x_{\text{NaCl satd.}} = 0.197$) to $\sim$40 MPa at 600°C ($x_{\text{NaCl satd.}} = 0.490$) [91]. Anderko and Pitzer [92] presented an EoS for the solubility of solid NaCl in SCW up to 900°C and 500 MPa. This approach has later been refined by Kosinsky and Anderko [93]. The EoS expresses the Helmholtz energy of the system at a given temperature, volume, and composition as the sum of the following terms, representing contributions from the ideal gas (^{ig}), repulsion (^{rep}), dipolar interactions (^{dip}), and a perturbation term (^{per}):

$$A(T, V, x) = A^{\text{ig}} + A^{\text{rep}} + A^{\text{dip}} + A^{\text{per}} \qquad (4.13)$$

The repulsion and perturbation terms are functions of η, proportional to the reduced density ρ_r

$$\eta = b(\omega)\rho_r/4 \qquad (4.14)$$

where $b(\omega)$ is the van der Waals covolume factor that, in turn, depends on the Pitzer acentric factor ω. Each of the terms in (4.13) is a multiparameter expression for the two components, water and NaCl (14 parameters for each), and their mixtures (32 parameters), as well as data for the melting of NaCl and its heat capacity and volume in the supercooled liquid state [93] and the dipole moment of water and the Na$^+$Cl$^-$ ion pair. The expressions are able to predict the solubility as well as the volumetric properties in this system up to 900°C and 500 MPa.

Tester and coworkers [94, 95] presented data on the solubility of sodium chloride and sulfate in steam at high temperatures as well as in SCW at conditions relevant to the SCWO process, 450–550°C and 10–25 MPa, as well as for the ternary systems. At these relatively low pressures the SCW has a low density (of the order of $100 \, \text{kg} \cdot \text{m}^{-3}$), hence is gas-like and may be called pressurized steam. The solubility of sodium chloride under such conditions is measured in ppm:

$$\log C_{\text{NaCl, ppm}} = 7.772 + 3.866 \log(\rho_W / \text{g} \cdot \text{cm}^{-3}) - 1233.4/(T/\text{K}) \quad (4.15)$$

The solubility as ppm corresponds to $\text{mg(NaCl)} \cdot (\text{kg SCW})^{-1}$ and can be converted to molality by division by the molar mass of NaCl: $58,450 \, \text{mg} \cdot \text{mol}^{-1}$ and the latter converted to mole fraction of salt by further division by $55.51 \, (\text{mol H}_2\text{O}) \cdot \text{kg}^{-1}$. Note the large difference between the solubility of NaCl in low density SCW, of the order of hundreds of ppm, Eq. (4.15), and the significant solubility at high density SCW, of mole fractions of some tenths, Eq. (4.12). The association to Na^+Cl^- ion pairs and the hydrolysis to $\text{NaOH} + \text{HCl}$ were considered by these authors as factors affecting the solubility. The coefficient 3.866 of $\log \rho_W$ represents the hydration number of the NaCl (see below).

Cui and Harris [96, 97] applied molecular dynamics simulation to the solubility of NaCl in SCW at relatively low pressures, up to 30 MPa [94], that correspond to gas-like densities between 30 and $150 \, \text{kg} \cdot \text{m}^{-3}$. The solubility is expressed as follows

$$RT \, x_{\text{NaCl satd.}} = \mu^{\text{cr}} - \mu^{\text{vap}} - \mu^{\text{ex}} \quad (4.16)$$

Here the μ's are chemical potentials at the states indicated by the superscripts: ^{cr} for the crystalline salt, ^{vap} for the bound ion pair in the vapor phase, and ^{ex} refers to the excess chemical potential for the salt in the solution. The first term was calculated by means of the quasiharmonic approximation, water is modeled by the SPC potential and 17 parameters describe the interactions between the O, H, Na, and Cl species in calculating the second term. One ion pair in 123 water molecules was used for the calculation of the third term. The results of the simulations: $s_{\text{NaCl}} = 156, 98$, and 51 ppm at 450, 500, and 550°C and 25 MPa, were in fair agreement with the experimental values [94].

The solubility of NaCl in low density SCW at 500°C and 27.5 MPa, 229 ppm according to Ref. [94], was confirmed as 227 ppm by Dell'Orco et al. [98], who proceeded to use their experimental method (precipitation from supersaturated solutions) to study the solubilities of LiNO_3, NaNO_3, and KNO_3 at 450–525°C and 24.8–30.2 MPa. Note that at these temperatures the pure alkali metal nitrates are molten rather than solid. At a given temperature the solubility of NaNO_3 increased with increasing pressure and density of the SCW: at 475°C from 372 ppm at 24.9 MPa to 1094 ppm at 30.3 MPa, and at a given pressure, ~27.5 MPa it decreased from 991 ppm at 450°C to 389 ppm at 525°C. The solubilities of LiNO_3 and KNO_3 at 475°C and ~27.5 MPa were

1175 and 402 ppm, compared with 630 ppm for $NaNO_3$ (all ppm values ± 20). The hydrolysis of the nitrates leads to formation of HNO_3 that in turn decomposes, so that after some time at SCW conditions nitrates are converted partly to nitrites, doing so increasingly at the higher temperatures.

Khan and Rogak [99] reported the solubility of Na_2CO_3 and Na_2SO_4 in SCW at pressures of ~ 25 MPa and temperatures from near the critical one to $438°C$ for the former salt and to $396°°C$ for the latter. At $383°C$ the solubilities were 1010 and 850 ppm for the two salts, decreasing steeply with increasing temperatures.

Leusbrock et al., in a series of papers [100–103] studied the solubility of various salts in SCW, but again at relatively low pressures, hence densities ≤ 200 kg·m^{-3}. They employed an aqueous solution of the salts, pressurized it to the target pressure and preheated it to SCW temperatures before letting it through an equilibration column and a filter to a space with reduced pressure and temperature for analysis of the salt content. In Ref. [100] they compared several approaches to the description of the solubilities. One is by means of an EoS, as discussed above concerning NaCl and the Anderko and Pitzer [92] EoS as refined by Kosinsky and Anderko [93]. Another approach using EoSs is that of Helgeson and coworkers [104, 105], resulting in a freely available software package for the calculations. These are multiparameter expressions, yielding beside the solubilities also such data as the partial molar enthalpies and volumes, but are rather cumbersome to use. Leusbrock et al. [100] preferred the semiempirical approach, based essentially on Eq. (4.1) (see also Eq. (4.15)), derived as follows. The solid salt $C_aA_c(s)$ dissolves to form the hydrated fully associated salt $C_aA_c \cdot nH_2O(f)$ in the fluid. The equilibrium constant K_s may be written on a molality basis as

$$K_s \approx m(C_aA_c \cdot nH_2O, f)/\rho_W{}^n \qquad (4.17)$$

using $\rho_W(T, P)$ to represent the activity of the solvent and ignoring activity coefficients of the solute, the molalities of the nonionized salt being quite low at low pressures. On further assumption of temperature-independent enthalpy $\Delta_{sln}H$ and entropy $\Delta_{sln}S$ of solution, Eq. (4.17) may be rewritten as

$$\log m(C_aA_c \cdot nH_2O, f) = -\Delta_{sln}H/\ln(10)RT + \Delta_{sln}S/\ln(10)R + n \log \rho_W$$
$$= a/T + b + n \log \rho_W$$

$$(4.18)$$

Over a limited temperature range the inclusion of a heat capacity of solution term $cT \ln T$ does not improve the description of experimental

TABLE 4.2   The Parameters for the Molal Solubilities of Salts in SCW, Eq. (4.17)

Salt	A	b	n	References
LiCl	-271	-3.79	2.48	3
LiNO$_3$	-815	-4.28	4.33	3
NaCl	-575	-4.94	4.57	100
NaCl	-981	-4.53	4.88	3
NaNO$_3$	-341	-4.11	3.06	100
NaNO$_3$	-269	-4.86	3.93	3
NaH$_2$PO$_4$	1859^a	-7.96	3.47	103
Na$_2$CO$_3$	453	-6.90	3.52	100
Na$_2$SO$_4$	-1637	-8.78	7.13	100
Na$_2$HPO$_4$	7439^a	-18.35	5.37	103
KOH	-698	-4.23	3.24	3
KCl	-685	-5.00	4.65	3
KNO$_3$	407^a	-5.85	3.72	3
KH$_2$PO$_4$	4418^a	-14.17	4.33	103
MgCl$_2$	0	-5.61	3.44	102
MgSO$_4$	431^a	-8.06	3.31	103
CaCl$_2$	-441	-4.32	2.52	102
CuO	-1241	-5.42	1.34	100
PbO	-1457	-3.03	1.97	100

^{a} It is not clear why these values are positive, signifying a negative enthalpy of solution, a feature that has not been discussed in the relevant publications.

data and its contribution may be included in the b parameter. The solubilities in SCW of NaCl (own experimental values at 380–410°C and 17–23 MPa [3]) and of NaNO$_3$, Na$_2$SO$_4$, Na$_2$CO$_3$, CuO, and PbO (from the literature) were described well by Eq. (4.18), with the parameters shown in Table 4.2. Leusbrock et al. also studied the solubilities of MgCl$_2$ and CaCl$_2$ in SCW at 387–397°C for the former and to 417°C for the latter salt and at 18.5–23.5 MPa, that is, again at low densities [102]. MgCl$_2$ was found to be hydrolyzed to a considerable extent under the experimental conditions used, causing Mg(OH)$_2$ to precipitate with evolution of HCl that resulted in a drop in the pH. After correction for the hydrolysis the solubility was described by Eq. (4.18) with the parameters shown in Table 4.2. CaCl$_2$ was only slightly hydrolyzed and no precipitation was found, although the pH also dropped somewhat. Corrections for the hydrolysis were applied and the results again conformed to Eq. (4.18) with the parameters shown in Table 4.2. Measurements on MgSO$_4$ and some alkali metal phosphate salts [103] at similar temperatures and pressures led again to results conforming to Eq. (4.18) shown in Table 4.2. The solubilities of KOH and K_2HPO_4 in SCW were measured between 423 and 525°C and between 400 and 450°C, respectively [106], and the results were expressed by means of Eq. (4.18), with the n values 3.0 and 8.8.

A review of mineral solubilities in SCW was reported by Walther [107]. The solubility of silica (quartz) in SCW, measured by several authors, was summarized by McKenzie and Helgeson [108].

4.3.2 Thermodynamic Properties

The thermodynamic and transport properties of solutions of salts (and electrolytes in general) in SCW are of cardinal importance in two areas. One is the behavior of supercritical hydrothermal fluids in geochemistry and the other is the applications of SCW, foremost being SCWO, that is, supercritical water oxidation, both discussed in Chapter 5.

Much of the available thermodynamic information has been summarized by Helgeson and coworkers in a series of papers. In an early paper [108] the permittivity of SCW was calculated from the solubility of silica and subsequently employed to estimate the thermodynamic properties of aqueous species to 900°C at 200 MPa. The data for ε_r have since been superseded by the more comprehensive and accurate values reported by Fernandez et al. [109]. The main point in the McKenzie–Helgeson treatment is the relation of the Gibbs energy or formation of the aqueous species at given T and P to the permittivity and its first and second temperature derivatives, according to the Born expression for ion hydration, by the specification of eight species-specific parameters.

In later papers Tanger and Helgeson [110] and Shock and Helgeson [104] presented a revision of the Helgeson–Kirkham–Flowers (HKF) equation of state [111] from which they calculated the standard thermodynamic and transport properties of aqueous species at high temperatures and pressures, including solutions in SCW. The standard thermodynamic properties of the ions are made up from solvation and nonsolvation contributions, both being in turn sums of intrinsic and electrostrictive contributions. As an example of the resulting expressions, the standard partial molar volume of an ion (or of an electrolyte, as a sum of ionic values) is given by

$$V_i^\infty = a_{1i} + a_{2i}(\Psi + P)^{-1} + [a_{3i} + a_{4i}((\Psi + P)^{-1})](T - \Theta)^{-1} - \omega_i Q$$
$$+ (\varepsilon_r^{-1} - 1)(d\omega_i/dP)_T$$

$$(4.19)$$

Here $\Psi = 260$ MPa and $\Theta = 228$ K are universal constants, $Q = (\partial \ln \varepsilon_r/\partial T)_P$, a_{1i}, a_{2i}, a_{3i}, and a_{4i} are temperature- and pressure-independent species-specific parameters, whereas $\omega_i = (N_A e^2/2)z_i^2/r_{ieff}$ represents the Born coefficient of a species, the effective radius of which, r_{ieff}, depends on the temperature and the pressure:

$$r_{ieff} = r_{ic} + 0.047(z_i + |z_i|) + |z_i|g \qquad (4.20)$$

Here r_{ic} is the crystal ionic radius (in nm), and the dependence is on both the absolute value and the sign of the ionic charge, and g is a solvent function of the pressure and the temperature [112]. The values of r_{ieff} decrease substantially in SCW at diminishing densities. The first two terms of Eq. (4.19) represent the nonsolvation contribution and the last two terms the Born solvation contribution.

Derivatives of Eq. (4.19) with respect to the pressure and the temperature yield the standard partial molar isothermal compressibility and expansibility, respectively. Similar expressions are derived for the heat capacity, entropy, enthalpy, and Gibbs energy from standard thermodynamic relationships, noting that $(\partial C_p/\partial P)_T = -T(\partial^2 V/\partial T^2)_P$. The nonsolvation contribution to the heat capacity becomes then $\Delta_{nonsol}C_P = c_1 + c_2(T - \Theta)^{-2}$, with c_1 and c_2 being species-specific constants. The enthalpy and Gibbs energy are complicated expressions relative to the values at the standard temperature of 298.15 K and pressure of 0.101325 MPa. The parameters $[a_1 + a_2(\Psi + P)^{-1}]$ and $[a_3 + a_4(\Psi + P)^{-1}]$ at 0.1 MPa (i.e., $a_1 + a_2/260$ and $a_3 + a_4/260$) and c_1 and c_2 for the nonsolvation contributions fitted to the temperature dependence of the standard partial molar volumes and heat capacities were tabulated [104] for many ions (monatomic as well as polyatomic) of importance to geochemical and other solutions of electrolytes in SCW. The solvation contribution depends through the effective radius on the function g [112] and on the permittivity and its pressure and temperature derivatives. Thus all the thermodynamic functions could be calculated up to 1000°C and 500 MPa for solutions of electrolytes in SCW from the expressions presented [104]. Small-scale figures showed calculated values for V^∞, C_P^∞, S^∞, and $\Delta_f H^\infty$ and $\Delta_f G^\infty$ of infinitely dilute Na^+, Cl^-, SO_4^{2-}, KOH, $CaCl_2$, and $MgSO_4$ at saturation pressures and in SCW at 100, 200, and 500 MPa as a function of the temperature up to 1000°C.

As an extension of this work, Oelkers et al. [113] presented standard partial molar Gibbs energy of formation, $\Delta_f G^\infty$, of a large number of aqueous species (inorganic and organic ions and neutral species) as well as minerals and gases over a wide range of temperatures and pressures, including those relevant to SCW up to 500 MPa and 1000°C. From these the equilibrium constants for chemical reactions in SCW can be estimated. To facilitate the calculations a freely available computer program, SUPCRT92, was provided.

TABLE 4.3   The Ion-Specific Parameters of Na^+ and OH^- for the Calculation of the Thermodynamic Properties of NaOH in SCW According to Shock and Helgeson's Equations

Parameter	Na^+	OH^-
$a_1/cm^3 \cdot mol^{-1}$	7.6944	5.2413
$a_2/cm^3 \cdot mol^{-1}$	−9560.44	308.779
$a_3/cm^3 \cdot (K \cdot mol^{-1})$	136.231	77.082
$a_4/cm^3 \cdot (K \cdot MPa^{-1} \cdot mol^{-1})$	-1.14056×10^5	-1.16403×10^5
$c_1/J \cdot (K^{-1} \cdot mol^{-1})$	76.07	17.36
$c_2/J \cdot (K \cdot mol^{-1})$	-1.2473×10^5	-4.3288×10^5

As an example for the application of these considerations, the thermodynamic functions for NaOH in SCW are shown in the following [114]. It should be noted that NaOH is miscible with water above its melting point, 319.1°C at sufficiently high pressures, such as the one that prevail in SCW (see Section 4.3.1). There are some experimental thermodynamic and transport data for NaOH in SCW for comparison: isochoric heat capacities C_V up to 550°C [115], densities up to 400°C and 400 MPa [87], and conductivities to 400°C and 300 MPa [89] and to 600°C and 300 MPa [116]. The crystal ionic radii employed are $r_c = 0.097$ nm for Na^+ and 0.140 nm for OH^-, and Eq. (4.20) yields the effective radius in terms of $g(T,P)$ [112]. The ion-specific parameters a_{ij} ($j = 1$–4) and c_{ij} ($j = 1, 2$) are shown in Table 4.3. The density and permittivity of the SCW are also needed, given in Chapter 2, Table 2.1. The resulting values of V^∞, $C_P{}^\infty$, S^∞, and $\Delta_f H^\infty$ and $\Delta_f G^\infty$ of NaOH in SCW are shown in Figs. 4.4–4.8. They may be compared with the very small scale figures for KOH reported by Shock and Helgeson [104].

4.3.3   Transport Properties

Although the titles of the papers by Helgeson and coworkers [104, 110] contain the words "transport properties" for aqueous species in SCW, such properties are not dealt with within these papers. Oelkers and Hekgeson corrected this in their later paper [117], based their considerations on experimental limiting conductivity data for electrolytes in SCW mostly up to 800°C and 400 MPa by Franck and coworkers, by Ho and Palmer, and by Marshall, Quist, and coworkers.

Electric conductivity data for electrolytes in SCW are mainly available from Ho and Palmer for HCl [118, 119], HBr [120], H_2SO_4 [121, 122], LiOH [123, 124], LiCl [123, 125], NaOH [116, 124], NaCl [125–127], NaBr [125, 128, 184], NaI [129], $NaCF_3SO_3$ [130], Na_2SO_4 [122], KCl [131] and KCl and KOH [132–134], $KHSO_4$ [135], KNO_3 [136], K_2SO_4 [137],

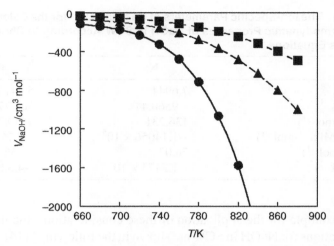

FIGURE 4.4 The standard partial molar volume of NaOH in SCW at 50 MPa (●), 70 MPa (▲), and 100 MPa (■) [114].

NH_4Br and $(CH_3)_4NBr$ [136], CsBr [125], and $MgCl_2$ and $CaCl_2$ [138]. The corresponding conductivities in SCW of the salts LiCl, NaI, KCl, KBr, KI, RbF, RbCl, RbBr, RbI, CsCl, CsBr, and CsI were added to this list by Quist and Marshall [139]. The values reported by Quist, Marshall, and coworkers are generally valid to 800°C and 400 MPa, those from Ho and Palmer are mainly up to 600°C and 300 MPa. The isobaric specific conductivities of dilute (0.01 molal) solutions of the alkali metal halides have maxima near 600°C.

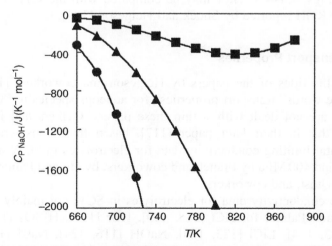

FIGURE 4.5 The standard partial molar isobaric heat capacity of NaOH in SCW at 50 MPa (●), 70 MPa (▲), and 100 MPa (■) [114].

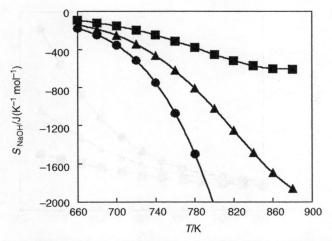

FIGURE 4.6 The standard partial molar entropy of NaOH in SCW at 50 MPa (•), 70 MPa (▲), and 100 MPa (■) [114].

The density dependence of the limiting molar conductivities Λ^∞_i of LiCl, NaCl, and KCl in SCW at various temperatures are shown in figures reported by Ibuki et al. [140]. The values of Λ^∞_i decrease essentially linearly with increasing ρ_W above $1.4\rho_c \sim 450\,\text{kg}\cdot\text{m}^{-3}$ as does the viscosity of the SCW.

Representative values of limiting ionic conductivities λ^∞_i (see below for their derivation) in SCW are shown in Table 4.4 at several temperatures and pressures. The values of λ^∞_i for singly charged ions, whether cations or anions

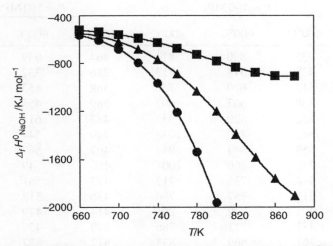

FIGURE 4.7 The standard partial molar enthalpy of formation of NaOH in SCW at 50 MPa (•), 70 MPa (▲), and 100 MPa (■) [114].

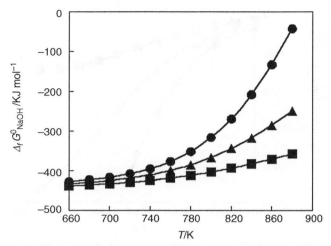

FIGURE 4.8   The standard partial molar Gibbs energy of formation of NaOH in SCW at 50 MPa (●), 70 MPa (▲), and 100 MPa (■) [114].

neither differ much among themselves (therefore, values for, say, Rb^+, Cs^+, and Ag^+ are not shown) nor do those of divalent cations (therefore, values for Sr^{2+}, Ba^{2+}, Mn^{2+}, Zn^{2+}, etc. are not shown). The hydrogen and hydroxide ions are obvious exceptions from these generalizations.

TABLE 4.4   Limiting Ionic Molar Conductivities in SCW, λ^∞_i/S·(cm²·mol⁻¹), at the Stated Pressures and Temperatures, Calculated from Reported [117] Tracer Diffusion Coefficients According to Eq. (4.23)

Ion	$P = 100$ MPa			$P = 500$ MPa		
	400°C	600°C	800°C	400°C	600°C	800°C
H^+	935	890	811	964	939	880
Li^+	428	579	713	286	332	331
Na^+	513	670	758	368	453	493
K^+	538	663	692	389	455	470
NH_4^+	632	829	935	483	618	718
Mg^{2+}	642	870	1038	449	544	585
Ca^{2+}	659	844	944	469	567	612
Fe^{2+}	539	860	1004	456	549	587
F^-	544	735	712	473	561	603
Cl^-	511	757	780	436	512	542
Br^-	494	729	801	414	479	498
I^-	481	711	786	399	458	468
OH^-	761	869	833	812	827	795
NO_3^-	426	579	598	333	337	285

Oelkers and Helgeson [117] deduced from the conductivity data and the previously established standard partial molar entropies of the electrolytes in SCW [104, 110] the linear relationship

$$\Lambda^\infty = a_j + b_j S^\infty \tag{4.21}$$

for a series of electrolytes of the same charge type sharing a common ion. The b_j coefficients are linear functions of the temperature and independent of the pressure.

In order to obtain the individual ionic conductivities, the limiting transference numbers are required. The ratio of the limiting transference numbers of Na^+ and Cl^- is known up to $125°C$, conforming to an Arrhenius expression:

$$t_{Cl^-}^\infty / t_{Na^+}^\infty = A_{NaCl} \exp(-E_{NaCl}/RT) \tag{4.22}$$

A logarithmic plot of this ratio within this range yielded the pre-exponential coefficient $A_{NaCl} = 0.969$ and an activation energy $-E_{NaCl} = 1107\,J \cdot mol^{-1}$ that were boldly assumed to be valid up to $800°C$. The ratio $t_{Cl^-}^\infty / t_{Na^+}^\infty$ thus varies from 0.795 at $400°C$ to 0.856 at $800°C$. Values of the limiting transference numbers of other salts involving sodium or chloride ions were then obtained from their corresponding limiting electrolyte conductivities, assuming transferability of ionic transference numbers from one salt to another (except for hydrogen and hydroxide ions).

Once the limiting equivalent conductivities of individual ions, $\lambda^\infty_i = t^\infty_i \Lambda^\infty_j$, have been calculated, they were converted to tracer diffusion coefficients of the ions in SCW according to the Nernst–Einstein relationship

$$D_i^\infty = (RT/F^2)|z_i|^{-1}\lambda_i^\infty \tag{4.23}$$

The calculated tracer diffusion coefficients D^∞_i of many mono- and divalent cations and a few anions in SCW at 100, 200, 300, 400, and 500 MPa at 400–1000°C were listed [117]. The values increase with the temperature but diminish with increasing pressure, except for the D^∞_i of H^+ and OH^- that are substantially independent of the pressure.

Ibuki et al. [141] analyzed the concentration dependence of the conductivity of uni-univalent electrolytes in SCW. They pointed out the necessity to carry out conductivity measurements at moderate concentrations, due to experimental constraints, hence the need for a reliable method for extrapolation to zero concentration. The fitting with the two parameters, Λ^∞_1 and K_{assoc} represented the conductivity data best. Ibuki et al. [140] also examined the applicability of the Hubbard–Onsager theory to the electrolyte conductivities in SCW and concluded that it does represent the experimental results satisfactorily at the higher densities, $\rho_W \geq 450\,kg\cdot m^{-3}$, better than the Walden product does. This is because the translational friction on a moving ion under these conditions is dominated by the dielectric friction, the solvent scaling length according to this theory being larger than the ionic radius, due to the small permittivity of the water.

Several authors applied MD computer simulations to the study of ion mobilities and conductivities in SCW. Balbuena et al. [142] used a semicontinuum model for the simulations and deduced values for the friction coefficient ζ^{-1} in terms of the reorientation times of water molecules in the bulk, $\tau_{W\,bulk}$, and in the first hydration shell of the ion, $\tau_{W\,ion}$. The resulting expression is

$$\zeta^{-1} = (\sigma\pi\eta_W r_i)^{-1}[(\tau_{W\,bulk}/\tau_{W\,ion}) + (r_i/r_{hyd})(1-(\tau_{W\,bulk}/\tau_{W\,ion})]\quad (4.24)$$

where $\sigma = 4$ is the stick boundary condition, η_W is the viscosity, r_i is the ionic radius, and r_{hyd} is the radius of the cavity, that is, of the ion with its first hydration shell, $r_{hyd} = r_i + 0.25 \pm 0.01$ nm. Given the friction coefficient, the limiting ionic conductivity is $\Lambda^\infty_i = Fe/\zeta$, the diffusion coefficient is $D^\infty_i = kT/\zeta$, and the Walden product is $\Lambda^\infty_i \eta_W = Fe\eta_W/\zeta$. The authors applied the SPC/E model for the MD simulations at 400°C to yield $\tau_{W\,bulk} = 0.19 \pm 0.02$ ps at densities from 87 to $500\,kg\cdot m^{-3}$ and at the density of $290\,kg\cdot m^{-3}$ $\tau_{W\,ion} = 1.07 \pm 0.11$ ps for Na^+, 0.56 ± 0.05 for K^+ and Rb^+, and 0.42 for Cl^-. Hyun et al. [143] considered at 400°C mainly low densities, and found in their MD simulations for Na^+ and Cl^- effective ionic radii that are inversely proportional to the Walden product and exhibit maxima at $\rho_W \sim 200\,kg\cdot m^{-3}$. At higher densities the first hydration shell depends only weakly on the density, but at lower densities there occurs a gradual loss of the primary hydration.

Lee et al. in a series of papers [144–149] reported MD studies of the ionic diffusion coefficients, ionic mobilities $u_i = D_i|z_i|F/RT$ and conductivities of ions and salts in SCW using the SPC/E water model. In the first paper [144] they considered Na^+ and Cl^- at 400°C and densities from 220 to $740\,kg\cdot m^{-3}$. The conductivity of NaCl is constant up to $\rho_W \sim 450\,kg\cdot m^{-3}$ and decreases linearly at higher densities in agreement with experiment (see Ibuki's results [140] above). In Ref. [145] Lee and Cummings dealt with LiCl, NaBr, and CsBr in a similar fashion. The change in slope of LiCl and CsBr conductivities occur at a lower density than for NaCl described above, the simulations agreeing with the experimental data. These authors [69] then applied the simulations to the virtual ions "Na^{2+}" and "Cl^{2-}" and the virtual neutral species "Na^0" and "Cl^0". The behavior of the virtual divalent ions shows the same trend as that of the univalent ones over the entire water density range tested. The diffusion coefficients of the virtual neutral species decrease monotonically with increasing densities attributed to the drastic increase in the hydration numbers of these species with the density. The SPC/E water model was abandoned in Ref. [147] in favor of a polarized water model, RPOL, applied to Li^+ in SCW, providing better agreement with experimental data. Both the SPC/E and RPOL models were then applied to the simulation of the conductivities of $CaCl_2$ and $MgCl_2$ [148, 149]. The divalent cations move with their first hydration shells and the interaction with the second shell restricts the mobilities of these ions.

4.3.4 Ion Association in SCW

The speciation of inorganic solutes in SCW is of great interest in geochemistry. The dissolution of minerals in hot brines (aqueous NaCl and HCl) is governed to a large extent by the concentration of free chloride ions, which is only a small fraction of the total chloride concentration [150].

The ionic dissociation of SCW itself has been discussed in Chapter 2, Section 2.5, and the pertinent equilibrium constants are listed in Table 2.9. The ionic product of water increases from the value $\log[K_W/(mol\cdot kg^{-1})^2] = -14$ at ambient conditions to considerably less negative values, such as -8.85 in SCW at 600°C and 500 MPa. There are, thus, $10^{-4.425}$ moles of hydrogen cations and hydroxide anions per kg of SCW under such conditions. On the other hand, contrary to the behavior of electrolytes in ambient water, where they are more or less completely dissociated into free ions whereas ion pairing is an exception, in SCW ionic dissociation is the exception, occurring generally only to a small extent and may in many cases be neglected altogether. This is certainly the case at gas-like densities of SCW (<200 $kg\cdot m^{-3}$) but may also be noted at liquid-like densities. Thus, ion association competes with ionic hydration under SCW conditions to determine the state of the species in electrolyte solutions in SCW.

The ion association of electrolytes in SCW is best studied by means of conductivity measurements as reported by many authors. The first who made precise electrolyte conductivity measurements on SCW and derived from them the ion pair association constants were Fogo et al. [151]. For many of the electrolytes studied the limiting conductivity in SCW Λ^∞ at ≥ 400°C and $\geq 600\ kg\cdot m^{-3}$ is nearly independent of the temperature but decreases with the density.

The association constant can often be fitted with four parameters as a function of the temperature and the density of the water:

$$\log K_{assoc} = a + b/T + (c + d/T)\log \rho_W \qquad (4.25)$$

Representative values of the parameters, valid for highly dilute solutions of the electrolyte in SCW, are shown in Table 4.5. Most, but not all, of these values were obtained from conductivity data, although the expressions used for obtaining the association constants varied from simple to more sophisticated ones. Some of the values were recalculated from dissociation constants (the reciprocals of the association constants) reported at various temperatures and density values as shown in Table 4.5, hence there are some large discrepancies in the parameters of Eq. (4.25). Other authors presented data in other forms, such as $\log K_{assoc(m)}(HCl) = 20.88 - 0.046P + 4.9 \times 10^{-5}P^2 - [12580 - 27.100P + 3.1 \times 10^{-2}P_2]/T$ with P at 50–500 MPa and T at 673–973 K [152], or as a power series in the square root of the density, as Liu et al. [153] did for $\log K_{assoc(c)}(NaCl)$. The subscripts $_{(c)}$ and $_{(m)}$ denote the pertinent concentration scale. The association of the ions in H_2SO_4 and Na_2SO_4 and their mixtures was studied by Hnedkovsky et al. [122].

Liu et al. [147, 153] applied quantum mechanical/molecular mechanical calculations to NaCl at 700°C and densities of 535, 280, 93.5, and 10.1 kg·m^{-3} and at 500°C at densities of 89.7 and 9.8 kg·m^{-3}. They obtained the Gibbs energy of hydration and the potential of mean force, from which the association constant was calculated, for instance $\log K_{assoc(c)} = 2.3 \pm 0.3$ at 700°C and 535 kg·m^{-3}, in reasonable agreement with experiment ($\log K_{assoc(c)} = 2.6$). Similar work was done by Cui and Harris earlier [96] and Liu et al. [153] compared the resulting values with experimental ones obtained by many authors, some of which are shown (as parameters of Eq. (4.25)) in Table 4.5. It should be noted that Eq. (4.25) is a fitting expression and its parameters have no direct physical significance, possibly except the coefficient of $\log \rho_W$.

Marshall among others interpreted the pressure and temperature dependence of the association constants [162]. He expressed the association constant of binary electrolytes in terms of the number n_W of water molecules released with the hydrated ions from the ion pair. On the molar scale

$$\log K_{assoc(c)} = \log K° + n_W \log(0.05551\rho_W) \tag{4.26}$$

where ρ_W is the density of water in kg·m^{-3}. In the case of NaCl at 500–800°C at pressures from 100 to 400 MPa $n_W = 9.7$ [162]. The expression $(c + d/T)$ of Eq. (4.25) may thus be equated with n_W.

The ion association of sodium and hydroxide ions in SCW to yield Na^+OH^- can be calculated [114] from the Bjerrum ion association theory, for which the required parameters are $a_B = r_{eff}(Na^+) + r_{eff}(OH^-)$ (see Eq. (4.20)) and the permittivity as functions of temperature and pressure. It turns out that the calculated association constant depends predominately on

TABLE 4.5 Parameters of the Association Constant of Electrolytes in SCW, Eq. (4.18) on the Molal Scale (Moles per kg SCW) or, Marked with an Asterisk, on the Molar Scale (Moles dm^{-3})

Electrolyte	$t/°C$	P/MPa or $\rho_W/kg \cdot m^{-3}$	a	b	c	d	References
HF*	450–650	ρ_W 300–700	30.26	−2,750	−7.28	0	154
HCl*	450–470	ρ_W 300–700	28.0	−1,986	−7.95	0	154
HCl	≤800	$P \leq 400$	5.405	−3,875	−16.93	0	155
HCl	380–410		28.31	−5,743	−6.05	0	156
HNO₃	380–400		24.35	832	−8.10	0	156
HNO₃	380–400	P 27.6–41.4	3.74	−1,530	−25.07	−11,540	157
H⁺HSO₄⁻	400		32.59		−10.50		158
H⁺HSO₄⁻	380–400	ρ_W 240–600	29.20	706	−9.60		156
LiCl*	450–650	ρ_W 300–700	37.16	−2,360	−11.62	0	154
LiCl*	400–900	ρ_W 700–1,000	12.87	−840	−3.45	0	159
LiCl	400–600	$P \leq 300$	0.724	−9	−15.80	5,431	123
LiCl	390–405	ρ_W 300–800	30.64	−10,774	−4.63	0	156
LiOH	400–600	$P \leq 300$	0.856	136	−15.00	4,226	123
LiOH	380–410	ρ_W 330–800	14.39	2,301	−5.58	0	156
NaOH	400–600	$P \leq 300$	1.65	−370	−16.22	6,300	116
NaOH	390–405	ρ_W 300–750	14.58	−975	−4.03	0	156
NaF*	550	ρ_W 300–700	29.09		−9.54		154
NaCl*	388	P 23–28	0.99		−10.23		151
NaCl*	400–800	ρ_W 300–750	1.21	−1,270	−10.2	0	120
NaCl	≤800	$P \leq 400$	1.197	−1,260	−12.20	0	155
NaCl	427, 527	P 25	4.50	−2,800	−10.2	0	160
NaCl	≤600	$P \leq 300$	0.997	−650	−13.42	2,600	127
NaCl	379–400	ρ_W 250–450	22.89	0	−7.74	0	125

(Continued)

NaCl	390–405	ρ_W 300–800	−6.02	18,233	−7.11	0	156
NaF$_3$CSO$_3$	≤450	ρ_W 300–800	0.888	330	15.83	5,349	130
KOH	400–600	$P \leq 300$	1.183	−133	−16.00	6,217	134
KOH	390–405	ρ_W 300–800	18.02	−1,917	−15.34	0	156
KCl*	400–700	ρ_W 600–900	13.39	−528	−3.61	0	161
KCl	400–600	$P \leq 300$	0.753	−100	−13.32	3,599	134
KCl	390–410	ρ_W 300–800	31.19	−10,367	−5.06	0	156
RbCl, CsCl*	450–650	300–700	29.42	−946	−9.45	0	154
CsBr	379–401	ρ_W 250–450	19.25	989	−6.99	0	125
NH$_4$OH*	380–600	ρ_W 500–850	33.3	0	−16.5	0	126
NH$_4$OH	400	ρ_W 500–900	50.26		−15.34		156
NH$_4$OH	400		4.24	0	−15.34	0	156
BaCl$^+$Cl$^-$*	420	ρ_W 500–1,000	19.95		−6.06		161

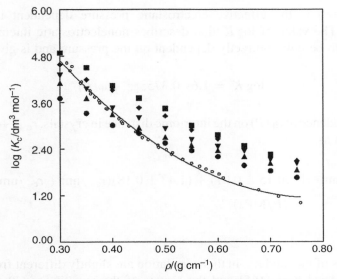

FIGURE 4.9   The calculated [114] logarithm of the association constant of Na$^+$ + OH$^-$ $\leftrightarrows$ Na$^+$OH$^-$ in SCW as a function of the density ($\bigcirc$) and fitted by Eq. (4.19a), upper line (—), and data by Ho and Palmer [116] at 400, 450, 500, 550, and 600°C (filled symbols from bottom up).

the density, rather than in detail on the temperature and pressure, and the resulting log $K_{assoc(c)}$ is shown in Fig. 4.9. This can be fitted by

$$\log(K_{assoc(c)}/dm^3 \cdot mol^{-1}) = 10.39 - 23.524\rho_W + 14.965\rho_W{}^2$$
$$= 1.658 - 386.4/T + (-14.336 + 6359/T)\log\rho_W$$
$$(4.27)$$

and is compared in Fig. 4.9 with data by Ho and Palmer [116]. Chialvo et al. [163] used MD simulations to obtain the potential of mean force and from it the association constant of very dilute Na$^+$Cl$^-$ associating at $T_r = 1.05$ and $\rho_r = 1.00$, namely log($K_{assoc(c)}$/dm$^3 \cdot$mol^{-1}) ranging from 3.71 to 4.36, depending on the water model employed. These values are comparable with that derived from the data of Ho and Palmer [116], 3.78.

Brady and Walther [164] used the Fuoss–Gilkerson expression rather than the Bjerrum one for the fitting of the association constants of various electrolytes obtained from conductivity measurements. Their expression for $K_{assoc(c)}$ on the molar scale can be rewritten as

$$\log K_{assoc(c)} = \log K' + 0.40 + 3\log a_F(P) + [7280z_+z_-/a_F(P)](T\varepsilon)^{-1}$$
$$(4.28)$$

where $a_F(P)$ is the effective electrostatic pressure dependent interionic distance. The value of $\log K'$ that describes nonelectrostatic interactions is assumed to be only indirectly dependent on the pressure and is given by

$$\log K' = 1.6 - 0.375 z_+ z_- / a_F(P) \qquad (4.29)$$

The dependence of $a_F(P)$ on the interionic distance in crystals, $r_{c+} + r_{c-}$, can be written as

$$a_F(P)/\text{nm} = [-(0.15 \pm 0.07) + (1.43 \pm 0.18)(r_{c+}/\text{nm} + r_{c-}/\text{nm})^{-1}]^{-1}$$
$$\times (P/\text{MPa})^{-1/2}$$

$$(4.30)$$

The values of r_{c+} and r_{c-} in this correlation are slightly different from those generally used here [165] and the scatter of the resulting $a_F(P)$ values is considerable. Still, association constant data for HCl, LiCl, NaCl, NaBr, NaI, KCl, $MgCl^+$, $CaCl^+$, and $H^+HSO_4^-$ could be modeled.

Ryzhenko and Bryzgalin [166] have previously modeled on an essentially similar electrostatic basis the association of acids in SCW: HF, HF_2^-, HNO_3, H^+HS^-, HS^-, HSO_4^-, $H^+HCO_3^-$, HCO_3^- among some others. The key variable determining the association constant, apart from the effective electrostatic interionic distance $d(P)$ that is pressure-dependent but not temperature-dependent, is the permittivity that depends on both thermodynamic conditions. Values of the permittivity of the SCW solvent are shown in Table 2.5, but solutions at appreciable concentrations of salts tend to have lower permittivities, hence the association should be enhanced relative to low solute concentrations.

4.3.5 Ion Hydration in SCW

The electric field exerted by ions attracts polar molecules to them: coordinate bonds between cations and (electron pair) donor atoms of the polar molecules and between anions and their hydrogen bond donor atoms are formed. This is the case in SCW as in common solvents and in water at ambient conditions. The main difference is the number of available water molecules in SCW, depending on its density, and the competition between the hydration of the ions, the association of oppositely charged ions discussed above in Section 4.3.2, and the self-association of the water molecules. As also for other properties of salts in SCW, sodium chloride has obtained the major attention earlier on.

Cummings and coworkers [34, 35] applied computer simulations with the SPC water model to obtain the pair correlation functions $g(I-W, r)$, where the ion I is Na^+ or Cl^- and W represents a water molecule. The excess of water molecules in a correlation region near an ion extending up to a distance R from its center is analogous to Eq. (4.2), with $g(I-W, r)$ replacing $g(Ar-W, r)$. At $T_r = 1.05$ and $\rho_r = 1.00$ a hydration number of 4 around each of the Na^+ and Cl^- ions resulted from the simulations and in the local region (extending to $R = 4\sigma_{OO}$, that is, four water molecule diameters) there were 26 and 20 water molecules, respectively, at a density manifold larger than the average bulk water density. At a denser state, $T_r = 1.00$ and $\rho_r = 1.50$, the excess water density is somewhat smaller. The average number of hydrogen bonds per water molecule, $\langle n_{HB} \rangle$, was found to be independent of the central particle, whether a water molecule or an ion, being near 0.8 at the lower density and near 1.0 at the higher one.

Cui and Harris [160] too applied the SPC model for water in their MD simulations to a cutoff $R = 1$ nm ($\sim 3.6\sigma_{OO}$), finding at a density of 300 kg·m^{-3} excess water molecules diminishing near Na^+ ions from 7.2 at 427°C to 4.4 at 727°C and near Cl^- ions from 10.1 to 4.9 over this temperature interval. It was difficult to assign hydration numbers to the ions on the basis of these numbers of excess water molecules, but they were 4.3 ± 0.4 for Na^+ and 7.1 ± 1.7 for Cl^-. Gupta and Johnston [167] calculated the mean hydration numbers of the sodium and chloride ions from a molecular thermodynamic model that included physical interactions (repulsion and dispersion forces) and hydrogen bonding. At a constant pressure of 30 MPa the hydration numbers diminish with increasing temperatures from 400 to 850°C as the density of the SCW decreases from ~ 350 to ~ 60 kg·m^{-3}: for Na^+ from 5.7 to 3.2 and for Cl^- from 5.4 to 1.9.

Bondarenko et al. [168] studied the structure of NaCl in SCW up to 500°C at the constant pressure of 100 MPa with the SPC/E water model and an NVT ensemble, that is, constant volume. The results (values at 400 and 500°C) are shown mainly in small-scale figures, so that quantitative conclusions are difficult to obtain from them. However, the strong hydrogen bonding from the water to the Cl^- are partly replaced by ion pairing, and same-charge association (e.g., Na^+-Na^+) is not excluded by the results.

Several authors subsequently studied the hydration in SCW of ions other than sodium and chloride. Flanagan et al. [169] studied the hydration structure of Li^+, Na^+, K^+, Be^{2+}, Mg^{2+}, Ca^{2+}, F^-, and Cl^- according to an adsorption model. They used MD simulations with the SPC/E water model at $T_r = 1.05$ and three densities 500, 290, and 87 kg·m^{-3}. The hydration numbers h_i at the lower density for this list of ions were 4.1, 4.8, 4.8, 4.0, 6.0, 8.0, 5.6, and 7.5, not significantly different from the values in ambient water. The hydration numbers were converted into local densities of the water by

$\rho^{loc} = h_i m_W / (4\pi/3)(r_{min}^3 - r_0^3)$, where m_W is the mass of a water molecule, r_0 is the radius of the bare ion, and r_{min} that of the edge of the hydration shell, given by the first minimum of $g(I-W, r)$, the ion–water pair correlation function. This local density is roughly equal to the maximal density that is, in turn, proportional to the surface charge densities of the ions. A concentric shell dielectric model was used for the calculation of the Helmholtz energies of hydration of the ions. The model consists of an inner hydration shell, where the density is given by $\rho^{loc} > \rho^{bulk}$, hence $\varepsilon^{loc} > \varepsilon^{bulk}$, and the bulk of SCW beyond this shell. The resulting Helmholtz energies of hydration of the ions are not very sensitive to the temperature and the density, provided the latter is $> 100 \, kg \cdot m^{-3}$. A compressible continuum model was also used with substantially similar results at the higher densities ($\geq 290 \, kg \cdot m^{-3}$), but is somewhat better at the lowest density tested ($87 \, kg \cdot m^{-3}$).

The compression of the SCW as a result of the electric field around a model ion (singly charged, in a cavity of 0.2 nm radius) was studied by Luo and Tucker [170] at three thermodynamic states of (T_r, ρ_r): (1.01, 0.77), (1.14, 0.875), and (1.30, 0.50). It was concluded that the compression-induced hydration Gibbs energy is largely determined within the first two hydration shells of the ion.

The thermodynamic quantities for hydration of the ions of sodium hydroxide can be calculated [114] from the heat capacity, entropy, enthalpy of formation, and Gibbs energy of formation, symbolized as Y^∞ in general, of the sodium and hydroxide ions in SCW (Section 4.3.2 and Figs. 4.4–4.8) and the ideal gas values for these quantities. The latter values, $Y^\circ(ig)$ in general, are known for 298.15 K and 0.1 MPa [165] and can be readily calculated for the conditions in SCW from the translational contribution to the entropy and the pressure-induced compression and temperature-induced expansion [171]. Hence,

$$\Delta_h Y^\infty(Na^+ + OH^-) = Y^\infty(NaOH, SCW) - [Y^\circ(Na^+, ig) + Y^\circ(OH^-, ig)]$$
$$(4.31)$$

and the values are shown in Table 4.6 for the isobaric transfer of the ions from the ideal gas state to SCW at 100 MPa.

The hydration of the ions and undissociated molecules of hydrogen chloride was studied by Mesmer et al. [155] experimentally and by Johnston et al. [172] according to the continuum electrostatic model. The hydration number change for $HCl \rightarrow H^+ + Cl^-$ obtained from Eq. (4.26) at a density of $800 \, kg \cdot m^{-3}$ diminish slightly (from 15 to 14) as the temperature is raised from 400 to 800°C. The numbers are larger at lower densities and much more strongly temperature dependent: at a density of $300 \, kg \cdot m^{-3}$ they decrease

TABLE 4.6   Standard Thermodynamic Quantities of Hydration of the Combined Ions of Sodium Hydroxide in Liquid Water at 0.1 MPa and 298 K and in Supercritical Water at 100 MPa at Various Temperatures

T/K	Δ_{hyd} $C_P°/J\,(K^{-1}\,mol^{-1})$	Δ_{hyd} $S°/J\,(K^{-1}\,mol^{-1})$	Δ_{hyd} $H°/kJ\,mol^{-1}$	Δ_{hyd} $G°/kJ\,mol^{-1}$
298	−126	−272	−936	−855
650	−94	−327	−1026	−813
700	−160	−401	−1072	−791
750	−316	−530	−1164	−767
800	−451	−705	−1298	−734
850	−435	−846	−1412	−693

from 315 to 21 over this temperature interval. What the significance of a hydration number of 315 is remains obscure [155]. The Helmholtz energy change for the iso-coulombic reaction

$$HCl + OH^- \leftrightarrows H_2O + Cl^- \qquad (4.32)$$

was calculated from the model [172] using thermodynamic cycles such as (2.38) for HCl as well as for H_2O. The solvation energies were calculated from the Born expression with specified cavity radii of the species that take electrostriction into account. The equilibrium (4.32) is shifted in SCW to the right, but by $\sim$63 kJ·mol^{-1} less than in the gas phase (where $\Delta_{equil}A = -234$ kJ·mol^{-1} for the Helmholtz energy), indicating that Cl$^-$ is more strongly hydrated than OH$^-$. The value of $\Delta_{equil}A$ becomes less negative by $\sim$8 kJ·mol^{-1} as the temperature is increased from 400 to 850°C at a density of 800 kg·m^{-3} and becomes more negative by $\sim$15 kJ·mol^{-1} as the density is decreased from 800 to 200 kg·m^{-3} at any temperature in this interval.

Fulton and coworkers [36, 173] used XAFS measurements to probe the near environments of Sr^{2+} and Rb$^+$ ions in SCW. These ions were chosen because of their suitability for XAFS measurements. In the case of Sr^{2+}, the states $T_r \approx 1.03$ and $P_r \approx 1.22$ ($\rho = 443$ kg·m^{-3}) and 1.53 ($\rho = 538$ kg·m^{-3}) were used and the $d_{Sr-O} = 0.258$ nm. The results showed 3.5 nearest neighbor water molecules at the lower density increasing to 3.8 neighbors at the higher one, considerably lower than under ambient conditions (7.3 neighbors). In the case of Rb$^+$, the states $T_r \approx 1.33$ and $P_r \approx 1.75$ ($\rho = 520$ kg·m^{-3}) and 2.86 ($\rho = 640$ kg·m^{-3}) were used, and $d_{Rb-O} = 0.281$ nm. The results were 3.5 nearest neighbors at the lower density increasing to 3.9 neighbors at the higher one, again lower under ambient conditions (5.6 neighbors). In both cases appreciable ion pairing should take place at the concentrations employed (0.5 mol·kg^{-1}), but the XAFS did not show this, beyond the concomitant reduction of the hydration number relative to its value under ambient conditions where no ion pairing occurs.

Yamaguchi et al. [174] studied the hydration of the ions of LiCl in SCHW (supercritical D_2O, $T_c = 370.7°C$) by neutron diffraction with isotope substitution, a method that permits the individual ion–water distances and coordination numbers to be determined. At $T_r = 1.03$ in a 5 molal LiCl solution, the distance $d_{Li-O} = 0.200$ nm and the coordination number 2.8 deuterium atoms around the chloride anion and at the same temperature and in 3 molal LiCl, the distance $d_{Cl-D} = 0.233$ nm with a coordination number of 2.5 were found.

4.4 BINARY MIXTURES OF COSOLVENTS WITH SCW

The possibility to tune the properties of SCW by changing the temperature and the pressure, hence, the density and the permittivity, makes SCW an almost "universal" solvent for all kinds of solutes: salts, polar and nonpolar organic compounds, gases, and so on. Therefore, there is little need to change the properties of SCW by means of a cosolvent, contrary, for instance to the case of supercritical carbon dioxide (SCD). The latter is an acceptable solvent for nonpolar organic compounds but polar ones (and salts) are not appreciably soluble in it, so that a cosolvent, such as methanol, is required for dissolving them.

Still, mixtures of SCW with SCD do arise in several processes: in SCW oxidation processes (SCWO, Chapter 5) CO_2 is produced by the oxidation of organic compounds and remains in solution in the SCW, thus eventually leading to mixtures of SCW and SCD. Geothermal processes occurring in deep layers of the earth also produce such mixtures. The solubility of CO_2 in SCW is dealt above in Section 4.1.1 and complete miscibility of the components occurs above the critical point of water.

The volumetric properties of the mixtures at 400°C and 10–100 MPa were determined by Seitz and Blencoe [175] and these, as well as the excess Gibbs energies were calculated for such mixtures and conditions (400°C and up to 400 MPa) by Blencoe et al. [176]. Some of the experimental densities of the mixtures are reproduced in Table 4.7, showing a minimum that occurs at increasing mole fractions of CO_2 as the pressure is increased. The excess volumes V^E are positive and have a maximum at 30 MPa. The $V^E(x_{CO_2})$ curves are skew (leaning toward the water end) at low pressures but become nearly symmetrical at high pressures. Franck and Tödheide [19] reported data at 400–750°C and 30–200 MPa for four mixtures, but in poor agreement with the more recent data at 400°C. The densities predicted from the EoS of Duan et al. [32] are also in only fair agreement with the experimental values at 400°C. However, this EoS predicted $PVTx$ data to 1000°C and 100 MPa that

TABLE 4.7   The Densities, ρ/kg·m^{-3}, of Mixtures of Water and Carbon Dioxide at 400°C at Various Pressures P/MPa and Mole Fractions of CO_2 from Ref. [175]

P/MPa	$x_{CO_2} = 0.1$	$x_{CO_2} = 0.3$	$x_{CO_2} = 0.5$	$x_{CO_2} = 0.7$	$x_{CO_2} = 0.9$
25	145	145	155	169	185
30	202	182	189	203	221
40	342	261	256	268	286
50	452	340	321	328	346
60	518	409	382	386	403
80	599	513	479	478	494
100	651	582	556	554	569

are in agreement with the data of Sterner and Bodnar [177] available up to 800°C and 600 MPa. As discussed earlier (in Section 4.1.1), the supercritical water + carbon dioxide mixtures become nearly ideal according to this EoS, which is above 700°C. Blencoe et al. [176] noted that the activity–composition predictions of Duan et al. [32] for 400°C differ from their own estimates, but concluded that revision of the fitting parameters could lead to improved predictions from that EoS.

Computer simulations have been applied to SCW + SCD mixtures by several authors. Destrigneville et al. [178] used Monte Carlo simulations for the mixtures at two isotherms: 500 and 800°C, pressures up to 3 GPa, and mole fractions $x_{CO_2} = 0.25, 0.40$, and 0.75, using the TIP4P potential model for the water and the MSM3 model for CO_2. The resulting $PVTx$ values lead to excess volumes, V^E, in good agreement with experimental values of Greenwood at these temperatures [30]. Lo Celso et al. [179] and Botti et al. [44] applied molecular dynamics simulations as well as neutron diffraction involving isotope substitution and empirical potential structure refinement computations to study mixtures of SCW and SCD. They employed the TIPS2 and EPM-M potential models for water and CO_2, respectively, at two concentrations: $x_{CO_2} = 0.0092$ and 0.092. The simulations reproduced the experimental site–site radial distribution functions adequately. The effect of the presence of the CO_2 is the reduction of the average number of hydrogen bonds formed by the water. This causes a diminution of the sizes of the water clusters and consequently the mixtures are below the percolation threshold (Section 3.1) [44].

This issue, of the local order induced in hot water by admixture of CO_2, has been tackled by infrared and Raman spectroscopy by Oparin et al. [180]. However, they just touched SCW, since the highest temperature studied at 25 MPa, 366°C, barely sufficed to produce total miscibility of the components as a supercritical fluid. It was concluded that water dimers are the predominant species at this condition with up to $x_{CO_2} = 0.10$. However, the insertion of hydrophobic CO_2 into the water structure diminishes the relative concentrations of the larger clusters (tri- and tetramers) compared with pure water. This is in agreement with the molecular dynamics and neutron scattering results presented above. The hot bands of the symmetrical stretching of CO_2 in the Raman spectrum of SCWO processes (Chapter 5) at 400–650°C are quite intense and separated from the fundamentals, and are temperature sensitive, so that they can be used for thermometry of such processes [45]. The ratio of the integrated band intensities of the hot band at 1409 cm^{-1}, I_B, and the fundamental at 1388 cm^{-1}, I_A, is linear with the temperature in the range tested, 390–540°C and $0.03 \leq x_{CO_2} \leq 0.20$. The expression

$$t/°C = 843.8(I_B/I_A) - 54.23 \qquad (4.33)$$

independent of the CO_2 content and the pressure, can then be used for the determination of the temperature of the mixtures.

Minami et al. [181] used UV-vis spectroscopy to study the hydrogen ion concentration in CO_2 solutions in SCW by using 4-nitrophenol as a probe (Section 4.2.2). They considered solutions at 380°C and a water density of 400 kg·m^{-3} with $0.03 \leq x_{CO_2} \leq 0.07$ and various NaOH concentrations. The reaction $2H_2O + CO_2 \leftrightharpoons H_3O^+ + HCO_3^-$ must be taken into account in such cases, with an equilibrium constant of 3.3×10^{-13}. This value is an order of magnitude smaller than the estimate from the revised HFK [111] EoS (Section 4.3.2.). The hydrogen ion concentration is 6–19 times higher in CO_2-containing SCW than in pure SCW (Section 2.5).

Geothermal processes produce fluids containing primarily H_2O and CO_2 from unstable hydrous and carbonate minerals [182]. The equilibrium quotient for their interaction has been estimated from the solubility of calcite ($CaCO_3$) in supercritical aqueous solutions containing CO_2 [107] as $10^{-7.2}$ at 500°C and 200 MPa, that is, several orders of magnitude larger than obtained with the spectroscopic probe at the lower temperature. However, this value was admitted to be a possible overestimate if carbonate complexes of calcium ($CaHCO_3^+$ or $CaCO_3$,aq) are formed. The solubility reaches a maximum as the mole fraction of CO_2 is increased: near $x_{CO_2} = 0.05$ at 400°C and nearer $x_{CO_2} = 0.02$ at 500 and 550°C at 200 MPa.

In an extension of the concept of "cosolvent," Ref. [183] can be cited in context to the delivery by methanol into an SCWO reactor (Section 5.2) of water-insoluble hazardous compounds, such as PCBs. The methanol (together with SCD) extracts the PCBs from contaminated soil, and participates in the oxidation process by being oxidized itself (serving as an energy-providing fuel) and as a catalyst. In the mixtures of hydrogen peroxide with SCW that are used for the oxidation, the H_2O_2 may also be considered as a reactive (and consumed) cosolvent.

4.5 SUMMARY

Except when used as a moderator/coolant in nuclear reactors, the SCW acts as a solvent as well as a reactant for some applications. Therefore, the ability of SCW to dissolve substances, as a function of the thermodynamic states of temperature and pressure that determine its density, is of prime importance. On the other hand, the properties of SCW are based on the readily available and cheap substance "water" that is environmental-friendly, nontoxic, and non-inflammable, in addition to its solvent properties being tunable, make SCW an ideal "green" solvent. This is true, except for the relatively harsh operation conditions required for the use of SCW: temperatures $\geq 374°C$ and pressures ≥ 22.5 MPa and for the concomitant corrosion problems of structural materials handling it.

The solubility of various solutes in SCW is dealt with within this chapter rather arbitrarily in terms of the solutes being gases, organic compounds, and salts. The solubility of gases is generally expressed as their Ostwald coefficients L_g, which are the ratios of the volumes of a pure nonreactive gas dissolving at a given temperature and pressure in a given volume of pure water, and the pressure is generally specified as 1 atm (0.101325 MPa). The mole fractions of dissolved gases in water at ambient conditions are generally $x_g < 1 \times 10^{-4}$, except for reactive gases, such as HCl, NH_3, and CO_2, as shown in Table 4.1. In SCW the solubilities are manifold larger and small molecular gases are completely miscible with SCW at sufficiently high temperatures. The critical phase boundary curves of binary systems involving SCW and many such gases have been reported; above the critical temperature surfaces $T_c(P, x_g)$ for these binary mixtures the two components are miscible. The solubilities of the gases in SCW are described by the three-dimensional critical bimodal surfaces. An isobaric projection of the $T_c-P_c-x_c$ surface for Ar in SCW is shown in Fig. 4.1 as an example. Below the surface, there are two fluid phases (liquid and gas) but above it there is a single fluid phase. SCW with the lightest gases H_2, He, and Ne exhibits gas–gas immiscibility. For the other gases a minimum in the critical temperature occurs. The system SCW $+ CO_2$ exhibits a huge decline in the two-phase critical point, the minimal critical temperature being $T_{c\,min} = 266°C$ at 245 MPa and $x_{CO_2} = 0.415$. The minimal critical temperatures for the binary mixtures of gases in SCW are shown in Table 4.1.

Water molecules in SCW may cluster around a solute gas molecule, but more commonly there is a deficit, and in the case of Ar of nearly three water molecules relative to their mean number around a water molecule (from MD simulations). A somewhat smaller deficit was obtained from an X-ray absorption fine structure study of Kr in SCW. The multiparameter Anderko–Pitzer equation of state (Section 2.1.3) was applied to solutions in SCW of the components of air, that is, oxygen and nitrogen, to obtain thermodynamic quantities pertaining to such solutions. Such solutes cause expansion of the SCW, hence relative deficits in the number of water molecules around them. This is also true for solutions of CO_2 in SCW, in which hydrogen bond percolation does no longer take place.

The phase equilibria between SCW and organic solutes that are not gaseous at ambient conditions have been mainly established for hydrocarbons: alkanes and aromatic ones. An example is shown in Fig. 4.2 for the SCW + hexane system. The limits $T_c(x)$ above which mixtures of SCW with hydrophilic solutes have complete miscibility are given by Eqs. (4.3) and (4.4). The solubility of solutes that are not miscible may be estimated from solubility parameter expressions. Cubic EoSs, Section 2.1.2, can be invoked for the components and the cross interaction parameters for binary mixtures are obtained from correlations with the known octanol/water partition coefficients.

Solubilities at ambient conditions are often estimated by means of the solvatochromic parameters of the solvent and the solute. In the case of solutions in SCW this is generally impossible, because of the instability of the commonly used solvatochromic indicators under the conditions prevailing in SCW. Still, the UV-vis spectra some compounds, such as benzophenone and aminobenzonitrile were measured in SCW in a static system up to 440°C and those of nitroanisole and nitroanilines in flow systems at up to 410°C. Clustering of water molecules near the polar solutes, interacting with their dipoles and by hydrogen bonding was observed. Such probes cannot, therefore, yield representative information regarding the bulk of the SCW.

The solubilities of ordinary salts in SCW at gas-like densities (≤ 200 $kg \cdot m^{-3}$) are greatly diminished compared to those in water at ambient conditions. The large lattice energies of crystalline salts are not sufficiently compensated by the hydration energies as fewer water molecules are available for the hydration of the ions. The low permittivity plays a role too, ion association being enhanced and the ion pairs are less well hydrated than the ions. As the density of the SCW is increased the solubility of many salts increases too, with hydration and ion pairing competing, both working against the lattice energy that must be invested. Mineral acids, such as HCl, HNO_3, H_2SO_4, and H_3PO_4, appear to be miscible with SCW at all proportions. Sodium hydroxide is completely miscible with water above its melting point of 319°C. The solubility of sodium chloride in SCW has attracted the major attention of investigators over many years. This, as also the solubilities s of other salts is commonly expressed in terms of Eq. (4.1): $\log s = k \log \rho_W + a/T + b$. Equation (4.18) provides an interpretation of the parameters, k (or n)

denoting the number of water molecules hydrating the ions, a denoting the enthalpy of solution, and b takes care of the entropy of solution and of the concentration scales and units used. Values of these parameters are shown in Table 4.2 for dissolution of many salts in SCW. The solubility of silica has obtained special attention in connection with mineral solubilities in SCW, but see Section 5.5.

Much of the available thermodynamic information regarding slats and minerals in SCW has been summarized by Helgeson and coworkers. The standard thermodynamic properties of the ions are made up from solvation and nonsolvation contributions, both being in turn sums of intrinsic and electrostrictive contributions. The standard partial molar volume of an ion is expressed in Eq. (4.19) as an example of the resulting expressions. These considerations are applied to the thermodynamic functions for NaOH in SCW shown in Figs. 4.4–4.8, with the ion-specific parameters shown in Table 4.3.

Electric conductivity data for electrolytes in SCW are mainly available from Ho and Palmer and from Quist and Marshall. Representative values of limiting ionic conductivities $\lambda^{\infty}{}_i$ in SCW are shown in Table 4.4 at several temperatures and pressures. The values of $\lambda^{\infty}{}_i$ for singly and doubly charged ions do not differ much among themselves for each charge type, whether cations or anions. Several authors applied MD computer simulations to the study of ion mobilities and conductivities in SCW. The ion association of electrolytes in SCW is best studied by means of conductivity measurements. The association constant can be fitted with four parameters as a function of the temperature and the density of the water: $\log K_{assoc} = a + b/T + (c + d/T) \log \rho_W$, Eq. (4.25). Representative values of the parameters are shown in Table 4.5. The electrostatic theories of Bjerrum and of Fuoss have been invoked to relate the values to the properties of the ions and SCW. The hydration of the ions, competing with their association has been studied extensively by MD computer simulations. The thermodynamic functions for the hydration of the ions of Na^+OH^- in SCW are shown in Table 4.6. The equilibrium $HCl + OH^- \leftrightarrows H_2O + Cl^-$ (4.32) is shifted in SCW to the right, indicating that Cl^- is more strongly hydrated than OH^-.

The possibility to tune the density, the permittivity, and other properties of SCW by changing the temperature and the pressure, obviates need to do so by means of a cosolvent. This is contrary to the case of SCD, which requires a cosolvent, such as methanol, to dissolve polar solutes. Still, mixtures of SCW with SCD do arise in several processes, such as in SCW oxidation (SCWO, Section 5.2), and in geothermal processes occurring in deep layers of the earth from unstable hydrous and carbonate minerals. Complete miscibility occurs above the $T_c(P, x)$ surface (Section 4.1.1). Some of the experimental densities of the mixtures are reproduced in Table 4.7, showing a minimum that occurs at increasing mole fractions of CO_2 as the pressure is increased. The reaction

$2H_2O + CO_2 \leftrightarrows H_3O^+ + HCO_3^-$ must be taken into account, the hydrogen ion concentration is 6–19 times higher than in pure SCW (Section 2.5).

REFERENCES

1. C. P. Casey, *Chem. Eng. News* **82**, 25 (2004).
2. K. Arai, R. L. Smith, and T. M. Aidam, *J. Supercrit. Fluids* **47**, 628 (2009).
3. A. Loppinet-Serani, C. Aymonier, and F. Cansell, *J. Chem. Technol. Biotechnol.* **85**, 583 (2009).
4. H. R. Patrick, K. Griffith, C. L. Liotta, C. A. Eckert, and R. Glaeser, *Ind. Eng. Chem. Res.* **40**, 6063 (2001).
5. E. Dinius and A. Kruse, *J. Phys. Condens. Mat.* **16**, S1161 (2004).
6. D. E. Knox, *Pure Appl. Chem.* **77**, 513 (2005).
7. R. Crovetto, R. Fernandez-Prini, and M. L. Japas, *J. Chem. Phys.* **76**, 1077 (1982).
8. R. Battino and H. L. Clever,in T. M. Letcher,Ed., *Developments and Applications in Solubility*, RSC Publishing, Cambridge, 2007, p. 66.
9. S. Cabani, P. Gianni, V. Mollica, and L. Lepori, *J. Solution Chem.* **10**, 563 (1981).
10. E. Wilhelm, R. Battino, and R. J. Wilcock, *Chem. Rev.* **77**, 219 (1977).
11. N. G. Sretenskaja, R. J. Sadus, and E. U. Franck, *J. Phys. Chem.* **99**, 4273 (1995).
12. A. E. Mather, R. J. Sadus, and E. U. Franck, *J. Chem. Thermodyn.* **25**, 771 (1993).
13. H. Lentz and E. U. Franck, *Ber. Bunsenges. Phys. Chem.* **73**, 28 (1969).
14. G. Wu, M. Heilig, H. Lentz, and E. U. Franck, *Ber. Bunsenges. Phys. Chem.* **94**, 24 (1990).
15. E. U. Franck, H. Lentz and H. Welsch, *Z. Phys. Chem. (NF)* **93**, 95 (1974).
16. T. M. Seward and E. U. Franck, *Ber. Bunsenges. Phys. Chem.* **85**, 2 (1981).
17. M. L. Japas and E. U. Franck, *Ber. Bunsenges. Phys. Chem.* **89**, 793 (1985).
18. M. L. Japas and E. U. Franck, *Ber. Bunsenges. Phys. Chem.* **89**, 1268 (1985).
19. E. U. Franck and K. Tödheide, *Z. Phys. Chem.* (Munich) **22**, 232 (1959).
20. K. Tödheide and E. U. Franck, *Z. Phys. Chem. (NF)* **37**, 388 (1963).
21. M. Christoforakos and E. U. Franck, *Ber. Bunsenges. Phys. Chem.* **90**, 780 (1986).
22. V. M. Shnonov, R. J. Sadus, and E. U. Franck, *J. Phys. Chem.* **97**, 9054 (1993).
23. A. Danell, K. Tödheide, and E. U. Franck, *Chem.-Ing.-Tech.* **39**, 816 (1967).
24. Y. S. Wei, R. J. Sadus, and E. U. Franck, *Fluid Phase Equil.* **123**, 1 (1996).
25. P. H. van Konynenburg and P. L. Scott, *Phil. Trans. R. Soc. (Lond.)* **298A**, 495 (1980).
26. M. Temkin, *Russ. J. Phys. Chem.* **33**, 275 (1959).

27. O. H. Scalise, *J. Chem. Phys.* **96**, 6958 (1992).

28. T. Yling, Th. Michelberger, and E. U. Franck *J. Chem. Thermodyn.* **23**, 105 (1991).

29. M. Neichel and E. U. Franck, *J. Supercrit. Fluids* **9**, 69 (1996).

30. H. J. Greenwood, *Am. J. Sci.* **273**, 561 (1973).

31. D. M. Kerrick and G. K. Jakobs, *Am. J. Sci.* **281**, 735 (1981).

32. Z. Duan, N. Møller, and J. H. Weare, *Geochim. Cosmochim. Acta* **56**, 2619 (1992).

33. N. G. Polikhrinidi, I. M. Abdulagatov, R. G. Batyrova, and G. V. Stepanov, *Intl. J. Refrig.* **32**, 1897 (2009).

34. P. T. Cummings, H. D. Cochran, J. M. Simonson, R. E. Mesmer, and S. Karaborni, *J. Chem. Phys.* **94**, 5606.

35. P. T. Cummings, H. D. Cochran, and S. Karaborni, *Fluid Phase Equil.* **71**, 1 (1992).

36. D. M. Pfund, J. G. Darab, J. L. Fulton, and Y. Ma, *J. Phys. Chem.* **98**, 13102 (1994).

37. S. F. Rice and J. J. Wickham, *J. Raman Spectrosc.* **31**, 619 (2000).

38. M. D. Bermejo, A. Martin, and M. J. Cocero, *J. Supercrit. Fluids* **42**, 27 (2007).

39. J. M. Seminario, M. C. Concha, J. S. Murray, and P. Politzer, *Chem. Phys. Lett.* **222**, 25 (1994).

40. T. Omori and Y. Kimura, *J. Chem. Phys.* **116**, 2680 (2002).

41. K. Sugimoto, H. Fujiwara, and S. Koda, *J. Supercrit. Fluids* **32**, 293 (2004).

42. K. Yui, H. Uchida, K. Itatani, and S. Koda, *Chem. Phys. Lett.* **477**, 85 (2009).

43. J. Duan, Y. Shim, and H. J. Kim, *J. Chem. Phys.* **124**, 204504 (2006).

44. A. Botti, F. Bruni, R. Mancinelli, M. A. Ricci, F. Lo Celso, R. Triolo, F. Ferrante, and A. K. Soper, *J. Chem. Phys.* **128**, 164504 (2008).

45. M. S. Brown and R. R. Steeper, *Appl. Spectrosc.* **45**, 1733 (1991).

46. N. Matubayashi and M Nakahara, *J. Chem. Phys.* **112**, 8089 (2000).

47. J. Hernandez-Cobos and L. F. Vega, *J. Mol. Liq.* **101**, 113 (2002).

48. Z. Alwani and G. M. Schneider, *Ber. Bunsengesel. Phys. Chem.*, **71**, 633 (1967).

49. Z. Alwani and G. M. Schneider, *Ber. Bunsengesel. Phys. Chem.* **73**, 294 (1969).

50. K. Bröllos, K. H. Peter, and G. M. Schneider, *Ber. Bunsenges. Phys. Chem.* **74**, 682 (1970).

51. R. Jockers, R. Paas, and G. M. Schneider, *Ber. Bunsengesel. Phys. Chem.* **81**, 1093 (1977).

52. R. Jockers and G. M. Schneider, *Ber. Bunsengesel. Phys. Chem.* **82**, 576 (1978).

53. Th. W. de Loos, W. G. Penders, and R. N. Lichtenthaler, *J. Chem. Thermodyn.* **14**, 83 (1982).

54. E. Brunner, *J. Chem. Thermodyn.* **22**, 335 (1990).

55. E. Brunner, M. C. Thies, and G. M. Schneider, *J. Supercrit. Fluids* **39**, 160 (2006).

56. I. M. Abdulagatov, A. R. Bazaev, R. K. Gasanov, and A. E. Ramazanova, *J. Chem. Thermodyn.* **28**, 1037 (1996).

57. I. M. Abdulagatov, A. R. Bazaev, R. K. Gasanov, E. A. Bazaev, and A. E. Ramazanova, *J. Supercrit. Fluids* **10**, 149 (1997).

58. S. M. Rasulov and I. M. Abdulagatov, *J. Chem. Eng. Data* **55**, 3247 (2010).

59. W. L. Marshall and E. V. Jones, *J. Inorg. Nucl. Chem.* **36**, 2319 (1974).

60. A. R. Bazaev, I. M. Abdulagatov. E. E. Bazaev, and A. Abdurashidova, *J. Chem. Thermodyn.* **39**, 385 (2007).

61. N. G. Polikhronidi, I. M. Abdulagatov, G. V. Stepanov, and R. G. Batyrova, *Fluid Phase Equi.* **252**, 33 (2007).

62. R. Biswas and B. Bagchi, *Proc. Indian Acad. Sci. (Chem. Sci.)* **111**, 387 (1999).

63. Y. Marcus, *Aust. J. Chem.* **36**, 1719 (1983).

64. Y. Marcus, *J. Supercrit. Fluids* **38**, 7 (2006).

65. S. Artemenko and V. Mazur, *Proc. 2004 ASME Intl. Mech. Eng. Congr.* 259 (2004); S. V. Artemenko and V. A.Mazur, in S. J. Rzoska and V. A. Mazur, Eds., *Soft Matter under Exoergic Impacts*, Springer, 2007.

66. G. E. Bennett and K. P. Johnston, *J. Phys. Chem.* **98**, 441 (1994).

67. T. L. Fonseca, H. C. George, K. Coutinho, and S Canuto, *J. Phys. Chem. A* **113**, 5112 (2009), see also T. L. Fonseca, K. Coutinho, and S. Canuto, *J. Chem. Phys.* **126**, 034508 (2007).

68. T. Xiang and K. P. Johnston, *J. Phys. Chem.* **98**, 7915 (1994).

69. J. Lu, J. S. Brown, C. L. Liotta, and C. A. Eckert, *Chem. Commun.* 665 (2001).

70. J. Lu, J. S. Brown, E. C. Boughner, C. L. Liotta, and C. A. Eckert, *Ind. Eng. Chem. Res.* **11**, 2835 (2002).

71. H. Oka and O. Kajimoto, *Phys. Chem. Chem. Phys.* **5**, 2535 (2003).

72. T. Fujisawa, M. Terazima, and Y. Kimura, *J. Phys. Chem. A* **112**, 5515 (2008).

73. K. Osawa, T. Fujisawa, M. Terazima, and Y. Kimura, *J. Phys. Conf. Series* **215**, 012092 (2010).

74. M. Osada, K. Toyoshima, T. Mizutani, K. Minjami, M. Watanabe, T. Adschiri, and K. Arai, *J. Chem. Phys.* **118**, 4573 (2003).

75. T. Aizawa, M. Kanakubo, Y. Ikushima, R. L. Smith, Jr., T. Saitoh, and N. Sugimoto, *Chem. Phys. Lett.* **393**, 31 (2004).

76. T. Aizawa, M. Kanakubo, Y. Hiejima, Y. Ikushima, and R. L. Smith, *J. Phys. Chem. A* **109**, 7375 (2005).

77. K. Minami, M. Mizuta, M. Suzuki, T. Aizawa, and K. Arai, *Phys. Chem. Chem. Phys.* **8**, 2257 (2006).

78. K. Minami, . Ohashi, M. Suzuki, T. Aizawa, T. Adschiri, and K. Arai, *Anal. Sci.* **22**, 1417 (2006).

79. K. Osawa, T. Hamamoto, T. Fujisawa, M. Terazima, H. Sato, and Y. Kimura, *J. Phys. Chem. A* **113**, 3143 (2009).

80. Y. Takebayashi, S. Yoda, T. Sugeta, K. Otake, T. Sako, and M. Nakahara, *J. Chem. Phys.* **120**, 6100 (2004).

81. T. L. Fonseca, K. Coutinho, and S. Canuto, *J. Chem. Phys.* **129**, 034502 (2008).

82. C. Nieto-Draghi, J. B. Avalos, O, Contreras, P. Ungerer, and J. Ridard, *J. Chem. Phys.* **121**, 10566 (2004).

83. M. Morimoto, S. Sato, and T. Takanohashi, *J. Jpn. Petrol. Inst.* **53**, 61 (2010).

84. V. M. Valyashko, in D. A. Palmer, R. Fernandez-Prini, and A. H. Havey, Eds., *Aqueous Systems at Elevated Temperatures and Pressures*, Chapter 15, p. 597, Elsevier, 2004.

85. V. Valyashko and M. Urisova, *Monatsh. Chem.* **134**, 679 (2003).

86. R. Cohen-Adad, A. Tranquard, R. Perrone, P. Negri, and A. P. Rollet, *Comput. Rend.* **C251**, 2035 (1960).

87. S. Kerschbaum and E. U. Franck, *Ber. Bunsenges. Phys. Chem.* **99**, 624 (1995).

88. M. R. Ally and J. Braunstein, *Fluid Phase Equil.* **87**, 213 (1993).

89. A. Eberz and E. U. Franck, *Ber. Bunsenges. Phys. Chem.* **99**, 1091 (1995).

90. J. L. Bischoff and K. S. Pitzer, *Am. J. Sci.* **289**, 217 (1989).

91. J. L. Bischoff, *Am. J. Sci.* **291**, 309 (1991).

92. A. Anderko and K. S. Pitzer, *Geochim. Cosmochim. Acta* **57**, 1657 (1993).

93. J. J. Kosinski and A. Anderko, *Fluid Phase Equil.* **183–184**, 75 (2001).

94. F. J. Armellini and J. W. Tester, *Fluid Phase Equil.* **84**, 123 (1993).

95. M. M. DiPippo, K. Sako, and J. W. Tester, *Fluid Phase Equil.* **157**, 229 (1999).

96. S. T. Cui and J. G. Harris, *J. Phys. Chem.* **99**, 2900 (1995).

97. S. T. Cui and J. G. Harris, *Intl. J. Thermophys.* **16**, 493 (1995).

98. P. Dell'Orco, H. Eaton, T. Reynolds, and S. Buelow, *J. Supercrit. Fluids* **8**, 217 (1995).

99. M. S. Khan and S. N. Rogak, *J. Supercrit. Fluids* **30**, 359 (2004).

100. I. Leusbrock, S. Metz, G. Rexwinkel, and G. F. Versteeg, *J. Supercrit. Fluids* **47**, 177 (2008).

101. I. Leusbrock, S. Metz, G. Rexwinkel, and G. F. Versteeg, *J. Chem. Eng. Data* **54**, 3215 (2009).

102. I. Leusbrock, S. Metz, G. Rexwinkel, and G. F. Versteeg, *J. Supercrit. Fluids* **53**, 17 (2010).

103. I. Leusbrock, S. Metz, G. Rexwinkel, and G. F. Versteeg, *J. Supercrit. Fluids* **54**, 1 (2010).

104. E. L. Shock and H. C. Helgeson, *Geochem. Cosmochem. Acta* **52**, 2009 (1988).

105. E. L. Shock and H. C. Helgeson, *Geochim. Cosmochim. Acta* **53**, 215 (1989).

106. W. T. Wofford, P. C. Dell'Orco, and E. Gloyna, *J. Chem. Eng. Data* **40**, 968 (1995).

107. J. V. Walther, *Pure Appl. Chem.* **58**, 1585 (1986).

108. W. F. McKenzie and H. C. Helgeson, *Geochem. Cosmochem. Acta* **48**, 2167 (1984).

109. D. P. Fernandez, A. R. H. Goodwin, E. W. Lemmon, J. M. H. Levelt Sengers, and R. C. Williams, *J. Phys. Chem. Ref. Data* **26**, 1125–1166 (1997).

110. J. C. Tanger and H. C. Helgeson, *Am. J. Sci.* **288**, 19 (1988).

111. H. C. Helgeson, D. H. Kirkham, and G. C. Flowers, *Am. J. Sci.* **281**, 1249 (1981).

112. E. L. Shock, E. H. Oelkers, D. A. Sverjensky, J. W. Johnson, and H. C. Helgeson, *J. Chem. Soc. Faraday Trans.* **88**, 803 (1992).

113. E. H. Oelkers, H. C. Helgeson, E. L. Shock, D. A. Sverjensky, J. W. Johnson, and V. A. Pokrovskii, *J. Phys. Chem. Ref. Data* **24**, 1401 (1995).

114. Y. Marcus, unpublished results, abstract at 6th Conf. on Supercrit. Fluids, Maiora, 2001.

115. I. M. Abdulagatov, V. I. Dvoryanchikov, B. A. Mursalov, and A. N. Kamalov, *Fluid Phase Equi.* **143**, m213 (1998).

116. P. C. Ho and D. A. Palmer, *J. Solution Chem.* **25**, 711 (1996).

117. E. H. Oelkers and H. C. Helgeson, *Geochem. Cosmochim. Acta* **52**, 63 (1996).

118. J. D. Frantz and W. L. Marshall, *Am. J. Sci.* **284**, 651 (1984).

119. P. C. Ho, D. A. Palmer, and M. S. Gruszkiewicz, *J. Phys. Chem. B* **105**, 1260 (2001).

120. A. S. Quist and W. L. Marshall, *J. Phys. Chem.* **72**, 684 (1968).

121. A. S. Quist, W. L. Marshall, and H. R. Jolley, *J. Phys. Chem.* **69**, 2726 (1965).

122. L. Hnedkovsky, R. H. Wood, and V. N. Balashov, *J. Phys. Chem. B* **109**, 9034 (2005).

123. P. C. Ho and D. A. Palmer, *J. Chem. Eng. Data* **43**, 162 (1998).

124. P. C. Ho and D. A. Palmer, *J. Phys. Chem. B* **104**, 12084 (2000).

125. M. S. Gruszkiewicz and T. H. Wood, *J. Phys. Chem. B* **101**, 6549 (1997).

126. A. S. Quist and W. L. Marshall, *J. Phys. Chem.* **72**, 3122 (1968).

127. P. C. Ho, D. A. Palmer, and R. E. Mesmer, *J. Solution Chem.* **23**, 997 (1994).

128. A. S. Quist and W. L. Marshall, *J. Phys. Chem.* **72**, 1545 (1968).

129. L. A. Dunn and W. L. Marshall, *J. Phys. Chem.* **73**, 723 (1969).

130. P. C. Ho and D. A. Palmer, *J. Solution Chem.* **24**, 753 (1995).

131. D. Hartmann and E. U. Franck, *Ber. Bunsenges. Phys. Chem.* **73**, 514 (1969).

132. E. U. Franck, *Z. Phys. Chem. (NF)* **8**, 92 (1956).

133. E. U. Franck, *Z. Phys. Chem. (NF)* **8**, 192 (1956).

134. P. C. Ho and D. A. Palmer, *Geochim. Cosmochim. Acta* **61**, 3027 (1997).

135. A. S. Quist and W. L. Marshall, *J. Phys. Chem.* **70**, 3714 (1966).

136. A. S. Quist and W. L. Marshall, *J. Chem. Eng. Data* **15**, 375 (1970).

137. A. S. Quist, E. U. Franck, H. R. Jolley, and W. L. Marshall, *J. Phys. Chem.* **67**, 2453 (1963).

138. J. D. Frantz and W. L. Marshall, *Am. J. Sci.* **282**, 1666 (1982).

139. A. S. Quist and W. L. Marshall, *J. Phys. Chem.* **73**, 978 (1969).

140. K. Ibuki, M. Ueno, and M. Nakahara, *J. Mol. Liquids* **98–99**, 129 (2002).

141. K. Ibuki, M. Ueno, and M. Nakahara, *J. Phys. Chem. B* **104**, 5139 (2000).

142. P. B. Balbuena, K. P. Johnston, P. J. Rossky, and J.-K. Hyun, *J. Phys. Chem. B* **102**, 3806 (1998).

143. J.-K. Hyun, K. P. Johnston, and P. J. Rossky, *J. Phys. Chem. B* **105**, 9302 (2001).

144. S. H. Lee, P. T. Cummings, J. M. Simonson, and R. E. Mesmer, *Chem. Phys. Lett.* **293**, 289 (1998).

145. S. H. Lee and P. T. Cummings, *J. Chem. Phys.* **112**, 864 (2000).

146. S. H. Lee and P. T. Cummings, *Mol. Simul.* **27**, 199 (2001).

147. S. H. Lee, *Mol. Simul.* **29**, 211 (2003).

148. S. H. Lee, *Mol. Simul.* **30**, 669 (2004).

149. G. H. Goo, G. Sungt, and S. H. Lee, *Mol. Simul.* **30**, 37 (2004).

150. H. P. Eugster, *Am. Miner.* **71**, 655 (1986).

151. J. K. Fogo, S. W. Benson, and C. S. Copeland, *J. Chem. Phys.* **22**, 212 (1954).

152. B. R. Tagirov, A. V. Zotov, and N. N. Akinfiev, *Geochim. Cosmochim. Acta* **61**, 4267 (1997).

153. W. Liu, R. H. Wood, and D. J. Doren, *J. Phys. Chem. B* **112**, 7289 (2008).

154. E. U. Franck, *Angew. Chem.* **73m**, 309 (1961).

155. R. E. Mesmer, W. L. Marshall, D. A. Palmer, J. M. Simonson, and H. F. Holmes, *J. Solution Chem.* **17**, 699 (1988).

156. K. Sue and K. Arai, *J. Supercrit. Fluids* **28**, 57 (2004).

157. K. J. Ziegler, L. Lasdon, J. Chlistunoff, and K. P. Johnston, *Comp. Chem.* **23**, 421 (1999).

158. T. Xiang, K. P. Johnstan, W. T. Wofford, and E. F. Gloyna, *Ind. Eng. Chem. Res.* **35**, 4788 (1996).

159. K. Mangold and E. U. Franck, *Ber. Bunsenges. Phys. Chem.* **73**, 21 (1969).

160. S. T. Cui and J. G. Harris, *Chem. Eng. Sci.* **49**, 2749 (1994).

161. G. Ritzert and E. U. Franck, *Ber. Bunsenges. Phys. Chem.* **72**, 798 (1968).

162. W. L. Marshall, *J. Phys. Chem.* **74**, 346 (1970).

163. A. A. Chialvo, P. T. Cummings, H. D. Cochran, J. M. Simonson, and R. E. Mesmer, *J. Chem. Phys.* **103**, 9379 (1995).

164. P. V. Brady and J. V. Walther, *Geochim. Cosmochim. Acta* **54**, 1555 (1990).

165. Y. Marcus, *Ion Properties*, Dekker, New York, 1997.

166. B. N. Ryzhenko and O. V. Bryzgalin, *Geokhimiya* **1987**, 137.

167. R. B. Gupta and K. P. Johnston, *Ind. Eng. Chem. Res.* **33**, 2819 (1994).

168. G. V. Bondarenko, Yu. E. Gorbaty, A. V. Okhulkov, and A. G. Kalinichev, *J. Phys. Chem. B* **110**, 4042 (2006).

169. L. W. Flanagan, P. B. Balbuena, K. P. Johnston, and P. J. Rossky, *J. Phys. Chem. B* **101**, 7998 (1997).

170. H. Luo and S. C. Tucker, *Theor. Chem. Acc.* **96**, 84 (1997).

171. A. Loewenschuss and Y. Marcus, *J. Phys. Chem. Ref. Data* **16**, 61 (1987).

172. K. P. Johnston, G. E. Bennett, P. B. Balbuena, and P. J. Rossky, *J. Am. Chem. Soc.* **118**, 6746 (1996).

173. J. L. Fulton, D. M. Pfund, S. L. Wallen, M. Newville, E. Q. A. Stern, and Y. Ma, *J. Chem. Phys* **105**, 2161 (1996).

174. T. Yamaguchi, H. Ohzono, M. Yamagami, K. Yoshida, and H. Wakita, *J. Mol. Liquids* **153**, 2 (2010).

175. J. C. Seitz and J. G. Blencoe, *Geochim. Cosmochim. Acta* **63**, 1559 (1999).

176. J. G. Blencoe, J. C. Seitz, and L. M. Anovitz, *Geochim. Cosmochim. Acta* **63**, 2393 (1999).

177. S. M. Sterner and R. J. Bodnar, *Am. J. Sci.* **291**, 1 (1991).

178. C. M. Destrigneville, J. P. Brodholt, and R. J. Wood, *Chem. Geol.* **133**, 53 (1996).

179. F. Lo Celso, R. Triolo, F. Ferrante, A. Botti, F. Bruni, R. Mancinelli, M. A. Ricci, and A. K. Soper, *J. Mol. Liq.* **136**, 294 (2007).

180. R. Oparin, T. Tassaign, Y. Danten, and M. Besnard, *J. Chem. Phys.* **123**, 224501 (2005).

181. K. Minami, T. Suzuki, T. Aizawa, K. Sue, K. Arai, and R. L. Smith, Jr., *Fluid Phase Equi.* **257**, 177 (2007).

182. J. V. Walther and P. M. Orville, *Contrib. Min. Petrol.* **79**, 252 (1982).

183. G. Anitescu and L. L. Tavlarides, *Ind. Eng. Chem. Res.* **41**, 9 (2002).

184. A. S. Quist and W. L. Marshall, *J. Phys. Chem.* **72**, 2100 (1968).

5

APPLICATIONS OF SCW

Supercritical water (SCW) has found so far many applications and more are being proposed all the time. These applications are based on several aspects of SCW, the main one being a "green" solvent, that is, being environment-friendly. Other aspects are its properties being tunable by changes of the temperature and pressure (Chapter 2) and its solvent power for a variety of substances (Chapter 4). Many reviews of actual and potential applications have been published, for example, recently by Brunner [1, 2] and Loppinet-Serani [3], among others.

The most important application of SCW is as a reaction medium, for example, in the conversion of biomass to hydrogen, natural gas, or liquid fuels, for total oxidation of obnoxious and hazardous materials (supercritical water oxidation (SCWO)), and in organic synthesis. Such applications include extraction, separation, and purification of materials. Other uses are in powder technology of inorganic substances and drug formulations, although the high temperatures involved severely limit the latter uses. SCW has been proposed as a coolant of nuclear reactors, allowing high thermodynamic efficiency. Some geothermal aspects of SCW should also be considered in this context.

The general conditions of the application of SCW are the temperature range of, say, 380–700°C and pressures of 25–40 MPa, although higher temperatures and pressures may also be used. Once a reaction or separation has been effected in a supercritical fluid, there are two general techniques for obtaining the desired product or for the disposal of unwanted waste: rapid expansion of supercritical solutions (RESS) and the addition of a supercritical antisolvent

Supercritical Water A Green Solvent: Properties and Uses, First Edition. Yizhak Marcus.
© 2012 John Wiley & Sons, Inc. Published 2012 by John Wiley & Sons, Inc.

(SAS). In case of SCW, where the recycling of the water is of minor importance, SAS technique is hardly used, but the RESS, with the concomitant cooling, precipitates the solutes if solid or produces a separate layer if liquid that can be handled and removed. Of course, the energy balance is a major item of concern as are the corrosive properties of SCW toward materials of construction of the reactors and pipes.

5.1 CONVERSION OF ORGANIC SUBSTANCES TO FUEL

One of the main uses envisioned for SCW is the conversion of biomass to biofuel. Biomass used for conversion to biofuel in SCW is derived from agricultural materials after their food values, if any, have been taken out: it consists mainly of cellulose (40–80%), hemicellulose (15–30%), and lignin (10–25%) [3]. It also contains other substances such as inorganic and organic compounds, the latter being tannins, terpenes, waxes, fatty acids, and proteins. A variant of this kind of processes is the conversion of wet biomass, for example, industrial wastewater from textile or food processing plants. A point that needs to be considered in the evaluation of such processes is whether the caloric value of the fuel that is obtained outweighs the caloric cost of heating and pressurizing the input materials to SCW states and maintaining such states for the duration of the process. An estimate of the break-even amount of energy involved has been made as $930 \, \text{kJ kg}^{-1}$ [4]. Of course, the cost of alternative means of disposal of the agricultural and industrial wastes must also be taken into account. Arai et al. [5] recently proposed a decentralized chemical process for dealing with biomass for a sustainable society.

In this section, the conversion in SCW of any organic material besides biomass to a combustible product, that is, to a fuel, is also being dealt with. Such processes include hydrolysis and thermal degradation, and are discussed under the two kinds of fuels produced: gaseous and liquid.

5.1.1 Conversion to Hydrogen and Natural Gas

The gas produced in an SCW gasification (SCWG) of biomass process is generally a mixture of hydrogen, methane, carbon monoxide, and carbon dioxide, called syn-gas. Small amounts of higher alkanes (ethane and propane) may also be produced. The hydrothermal SCWG process has the advantage that biomass with a large water content (up to 80 mass%) can be processed [6]. This water then constitutes the supercritical solvent and reaction medium. The carbon dioxide produced dissolves readily in SCW, letting the hydrogen and methane to be collected as gases (Section 4.1). Furthermore, the water gas shift reaction

$$H_2O + CO \rightarrow H_2 + CO_2 \tag{5.1}$$

may be invoked to get rid of the carbon monoxide and produce more hydrogen.

According to the operation conditions, hydrogen or methane is the main product [7, 8]. A catalyst is useful for the reduction of the production of undesirable tar and coke by-products. At the lower temperature range of SCW, near 400°C, methane is the main gaseous product, but at higher temperatures (near 650°C) hydrogen becomes the major combustible product. However, as the feed concentration is increased (up to 30 mass%), the yield of hydrogen diminishes and that of methane increases [9]. An activated nickel catalyst increases the yield of methane near 400°C in the gasification of wet biomass [7]. For the production of hydrogen, a process not involving a heterogeneous catalyst is preferred, since plugging by salts resulting from inorganic compounds present in the biomass could then be avoided.

A demonstration plant VERENA was constructed in Karlsruhe, Germany, with a throughput of $100 \, kg \, h^{-1}$ biomass slurry with up to 20 mass% dry matter (Fig. 5.1), as described by Kruse [9]. The biomass is crushed, its water content is adjusted to the desired value, and the slurry is pressurized and preheated in a heat exchanger to just below the critical point of water. It is then fed into the reactor, made of a nickel alloy, and the pressure is increased to 35 MPa and the temperature to 700°C, at which conditions SCWG process takes place. The mixture is then cooled under pressure, and the combustible product gases, hydrogen and methane, are obtained at a high pressure (20 MPa) to be used as fuel gases. Further cooling and release of the pressure permit the dissolved carbon dioxide to be vented. It was found that the alkaline salts in the reactor brine catalyze the water gas shift reaction (5.1). The formation of "active hydrogen" during the process is apparently useful for the prevention of polymerization of intermediates leading to tar formation. The merits of and problems with this kind of SCWG processes were discussed by Gasafi et al. [10], where the feed was sewage sludge rather than agricultural biomass.

The kinetics of the water gas shift reaction in SCW in the absence of a catalyst was studied by Sato et al. [11]. They showed that since water is not only the solvent but also a reactant (5.1), its density, as determined by the pressure, is an important variable. The resulting first-order rate constant for the noncatalyzed CO conversion at 380–590°C and 10–60 MPa is

$$\log k = (5.58 \pm 1.38) - (5.0 \pm 0.8) \times 10^4/RT \tag{5.2}$$

A thermodynamic equilibrium analysis of the SCWG process was undertaken by Tang and Kitagawa [12] and by Letellier et al. [13]. Both groups

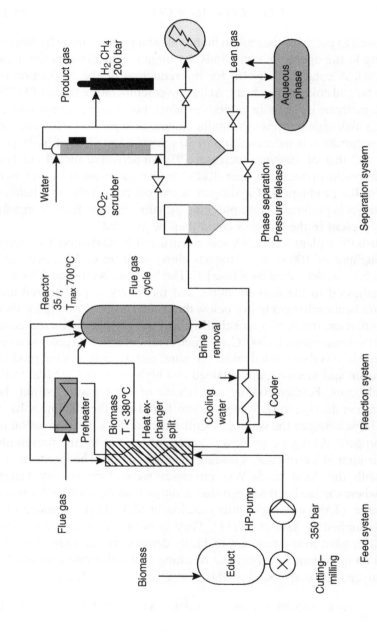

FIGURE 5.1 Schematics of the VERENA demonstration plant. Adapted with permission from Ref. [9].

Product gas

H₂ CH₄
200 bar

Lean gas

Aqueous
phase

Water

CO₂-
scrubber

Phase separation
Pressure release

Separation system

Reactor
35 *l*,
T_max 700°C

Flue gas
cycle

Brine removal

Preheater

Biomass
T < 380°C

Heat ex-
changer
split

Cooling
water

Cooler

Reaction system

Flue gas

HP-pump

Biomass

Educt

Cutting—
milling

350 bar

Feed system

found it expedient to use the Peng–Robinson equation of states (EoS) for the multitude of components existing in the process, because it requires fewer parameters than, say, the Duan EoS (Section 2.1.2). Letellier et al. [13] modeled the process as consisting of a reactor and a separator in which chemical equilibria take place and considered 13 relevant species: H_2O, H_2, CO, CO_2, CH_4, O_2, NH_3, H_2S, CH_3CHO, CH_3CO_2H, C_6H_5OH, C(s), and INO (the latter does not undergo chemical reactions). The INO, ammonia, and hydrogen sulfide result from inorganic components, nitrogen, and sulfur atoms in the compounds making up the biomass. The three liquid organic compounds, acetaldehyde, acetic acid, and phenol, are possible products from incomplete gasification, as is the solid carbon C(s). Mass and atom balances are taken into account in the model, as are the seven equilibria in addition to (5.1), namely, reactions of $C(s) + O_2 \rightarrow CO_2$, $C(s) + 2H_2 \rightarrow CH_4$, and $C(s) + CO_2 \rightarrow 2CO$ and hydrolysis of the three liquid organic compounds to form H_2 and CO. The activities of the species in SCW are then calculated from the pressure and the fugacity coefficients, the latter from the said EoS by means of the critical properties of the species.

This model was able to predict the composition of the output gases in the SCWG process for glucose (representing cellulose) at 10 mass% in water, studied experimentally by Lee et al. [14], only at the highest temperature considered, 740°C [13]. At lower temperatures the model fails, apparently because equilibrium is not achieved due to slow kinetics.

The phase equilibria involved in the SCWG process were studied by Feng et al. [15], mainly for the removal of CO_2 and the production of higher purity hydrogen. The reaction mixture from the reactor at 600°C and 35 MPa is retained at this pressure but cooled to 25°C, under which conditions most of the CO_2 remains dissolved in the water, but the H_2 is released. The solubilities of the components of the syn-gas are modeled with several EoSs (Section 2.1.2). About 75% of the hydrogen is thus recovered in the gas phase, the remaining being in the liquid, and is recovered by lowering the pressure.

Major applications of the SCWG process may be envisioned, dealing with industrial waste streams. A fairly early attempt to deal with sewage was reported by Schmieder et al. [6], where a hydrogen-rich gas resulted from SCWG at 450°C in the presence of potassium hydroxide or carbonate with a very dilute feed. A more comprehensive treatment of real waste biomass in SCW was reported by Jarana et al. [16], who mentioned previous works by Antal and coworkers [17] and Guo et al. [18]. They dealt with industrial wastes containing a high concentration of organic matter: cutting oil wastes and vinasses from alcohol distilleries. The operating conditions were ≥450°C and 25 MPa and a concentration of the feed expressed as $COD = 12\,g\,dm^{-3}$. Some of the experiments were carried out in the absence of oxygen, whereas others in its presence at 10% of the chemical oxygen demand (COD);

then again some experiments were carried out in the absence of a catalyst, whereas others in the presence of $0.004\,mol\,dm^{-3}$ (0.02 mass%) KOH [16]. The presence of the oxygen, the alkali, and a higher temperature (550°C) were beneficial for hydrogen formation from both kinds of waste feeds, but more so for the vinasses. Larger amounts of oxygen, however, diminished the hydrogen yield.

A different kind of catalyst, namely, nickel on zirconia, was used for the SCWG process to obtain hydrogen from wastewater containing polyethylene glycol (PEG) [19]. The operation conditions were 390°C and 24 MPa and $2\,g\,dm^{-3}$ PEG feed. Optimal amounts of catalyst and residence times in the continuous flow reactor were studied, the gasification efficiency increased with both parameters, and continuous operation for 85 h was demonstrated. A Turkish group [20, 21] studied the SCWG process for olive mill and textile wastewaters. The former has a high content of polyphenols as a distinguishing feature. The operating conditions were 25 MPa and 400–600°C and the process variables were the residence times in the reactor. Characteristic total organic carbon (TOC) contents of $6.1\,g\,dm^{-3}$ of the olive mill wastewater and $0.77\,g\,dm^{-3}$ of the textile wastewater were treated, the latter with $3.5\,g\,dm^{-3}$ COD. Gaseous C_1–C_4 products accompanied hydrogen and carbon dioxide in this process.

5.1.2 Conversion to Liquid Fuel

Ethanol is a desirable liquid fuel obtainable from plant material (lignocellulose), indirectly, by an SCW process. Although biomass as plant material consists mainly of cellulose, it contains some 10–25% lignin, a hydroxyaromatic polymer. This prevents efficient enzymatic attack of the cellulose, so that the latter cannot be readily fermented to ethanol. SCW pretreatment, catalyzed by carbon dioxide acting as an acid, hydrolyzes the polymeric hydroxyaromatic substances to a fermentable broth [22, 23]. Mild SCW conditions 400°C and 25 MPa are used in order to avoid further degradation of the water-soluble oligosaccharides to be fermented in a second stage. Saka and coworkers [22, 24–26] studied this process in detail in order to find the most efficient way to obtain the fermentable broth and high ethanol yields. A flow-type reactor produced larger yields of water-soluble hydrolyzed products and fewer pyrolyzed ones than a batch process [24]. Treatment of the water-soluble product with wood charcoal to remove fermentation-inhibiting furan and phenolic products was essential for obtaining high ethanol yields [25]. Alkaline treatment was also effective for this purpose [22]. The kinetics of the cellulose conversion in SCW (400°C and 25 MPa) to water-soluble products was studied by Sasaki et al. [27], who found a homogeneous first-order rate law.

A different route to liquid fuels by SCW processing is the degradation of waste plastic materials, such as hydrocarbons (polyethylene (PE)), esters (polyethylene terephthalate (PET)), and amides (nylon) [28, 29]. Pyrolysis of PE in SCW at 420°C and ~35 MPa produced alkanes and alkenes in the C_2–C_9 range [29]. The initial pyrolysis is heterogeneous with a molten PE phase, but the initial products at this temperature (e.g., hexatriacontane, C_{36}) are miscible with the SCW and their homogeneous degradation proceeds faster. No actual recovery of liquid alkane fuel, however, was discussed.

Bitumen, whether an undesirable oil refinery by-product (as vacuum residue) or as obtained from oil sands or oil shales, is an important prospective source for liquid fuels. Its extraction from low-lying oil sand layers is carried out with the steam-assisted gravity drainage (SAGD) process, so that pre-heating and mixing with water have already been effected for subsequent processing by SCW. The latter constitutes not only a solvent but also a reactant that supplies the hydrogen atoms needed to cap radical intermediates, thus preventing coke production. Studies on such processes have been undertaken mainly by Chinese and Japanese groups [21, 30–34].

Zhao et al. [31] and Cheng et al. [32] treated vacuum residue (heavy oil fraction in refinery) in a batch process at 420°C and 25 MPa and obtained light oils (boiling below 350°C) in 83% yield having an average molar mass of 35% and a viscosity of 0.6% of those of the feed. Also, the coke formation was reduced to 3.6 mass% compared to 14.9% in thermolysis in the absence of SCW.

Hu et al. [30] studied the extraction of Chinese (Huadian) oil shale with SCW and obtained 57% an extract yield at 400°C and 25–30 MPa, consisting of some 50 mass% oil (the rest is asphaltene). Morimoto et al. [33] studied the effect of SCW at 429–450°C and 20–30 MPa on the upgrading reaction of oil sand bitumen. They concluded that the main effect of SCW is to disperse the extracted bitumen and permit its upgrading to middle distillates, rather than to provide hydrogen atoms for radical capping. Sasaki et al. [34] studied the bitumen liquefaction at 400°C and concluded that the process was accelerated with increasing water density (pressure). Sato et al. [35] studied the upgrading of bitumen obtained from the SAGD process in SCW at 400–480°C and found that the presence of formic acid has a beneficial effect in that higher conversion of asphaltene and lower coke formation were achieved. The optimal temperature was 450°C, since at higher ones coke formation increased.

5.2 SUPERCRITICAL WATER OXIDATION

Supercritical water oxidation has been a "hot' issue and promises to remain so, that is, a principal application of SCW. Recent review papers demonstrate this [2, 36, 37]. SCWO process aims to destroy hazardous and noxious organic substances and convert them to innocuous, environment-friendly products. The fundamental aspects of the application rest on the good solubility of organic substances and oxygen in SCW at the typical operation conditions: 650°C and 25 MPa. Therefore, the oxidation reaction occurs in a homogeneous phase without problems of mass transfer and interface crossing. Another feature of consequence is the heat generated in the exothermic oxidation process, which if properly handled is able to sustain all the energy requirements of the process and even leave some energy for external use.

Examples of the materials to be treated are toxic or noxious ones, such as dioxins, pesticides, contaminated soils, biowastes, industrial wastewater, sewage sludge, and so on. The products of the process are carbon dioxide, water vapor, and nitrogen (mainly if air is used as the oxidant), vented as gases, and hydrochloric, sulfuric, and phosphoric acids, absorbed in a base, such as sodium hydroxide or carbonate, the salts being disposed of as a concentrated brine. Nitrogenous organic substances may be oxidized to NO_x that must also be absorbed in the base. Important issues that need consideration are the corrosion of the construction materials of the reactor and pipes, and so on (dealt with in Section 5.7), the avoidance of plugging of the reactor and piping by scale due to deposited salts and their ultimate disposal as brines, and the energy balance of the process. Delayed solutions of such problems, and the economic aspects, have slowed down the commercialization of the SCWO process.

As of 2007, several industrial size SCWO plants were operative, mainly in Japan and the United States and also in the United Kingdom and in Korea [37]. Several demonstration and pilot plants have been operative elsewhere too, for example, in Germany and Spain. The U.S. armed forces operate several SCWO plants for the total destruction of chemical warfare agents, propellants, smokes, and so on.

5.2.1 General Aspects of SCWO Process

The SCWO process consists of four main steps: feed preparation, reaction, salt separation, and heat recovery, as shown in Fig. 5.2 [36].

The feed is characterized by its total organic carbon content in mass% or mass per unit volume or by its chemical oxygen demand for total oxidation (in mass of oxygen per unit mass or volume of the feed). The feed preparation includes the choice of the oxidant: oxygen as air, liquid oxygen, or hydrogen peroxide, dictated by the economics or even produced *in situ* by electrolysis [38]. The efficiency of the process does not depend on this choice [39]. The material to be oxidized, as an aqueous slurry, and the oxidant are fed through a pressurizing pump into the reactor, where the exergonic oxidation

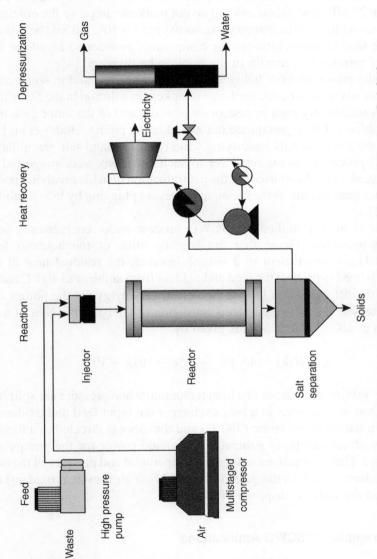

FIGURE 5.2 Schematics of an SCWO plant. Adapted with permission from Ref. [36].

159

takes place, the heat produced being sufficient to keep the reaction mixture at the required temperature. There is a play-off between the temperature of the process and the residence time: complete oxidation with stoichiometric amounts of the oxidant occurs at 650°C at a residence time < 50 s. Higher pressures than 25 MPa and excess oxidant do not markedly improve the efficiency of the process [36]. On the contrary, excess oxygen (> 10%) should be avoided when the feed contains nitrogenous compounds in order to avoid the formation of nitrates that remain in the brine to be disposed [4].

The salts produced from halogen, sulfur, phosphorus, and in some cases nitrogen atoms in the organic feed are either kept in solution in the SCW if its density is sufficiently high or else in suspension, and in the latter case may cause problems if they precipitate out and plug the piping. Hodes et al. [40] reported the fundamentals underlying scale formation and salt precipitation in SCWO processes. Some remedies to such situations were suggested by Marrone et al. [41]. Bases used for the neutralization of acids involving potassium rather than sodium as the cation may prevent plugging by low-solubility NaCl [42].

Cocero et al. [4] studied the SCWO process under energetically self-sufficient operation. Given that the heating value of the aqueous feed is $\geq 930 \, \mathrm{kJ \, kg^{-1}}$ (equivalent to 2 mass% hexane), the residual heat of the products is used to preheat the feed and oxidant from ambient to 400°C and to generate (by off-gases in a turbine) the electrical energy for the pumps and air compressor. For an organic feed consisting of $C_c H_h O_o$, the heat of total oxidation at 400°C and 25 MPa is given by

$$\Delta_{\mathrm{react}} H(\mathrm{kJ \, mol^{-1}}) = -415c - 107h + 193o \qquad (5.3)$$

The gases leaving the reactor at a high temperature and pressure are split into two streams: one preheats in a heat exchanger the input feed and oxidant to near the initiation temperature (380°C) and the other is directed to a turbine coupled with an electricity generator to provide power for the pumps and compressor. These operations reduce the temperature and pressure of the duct to near ambient, at which the gases (CO_2 and N_2, if air is used as oxidant) are vented and the water is disposed of [36].

5.2.2 Examples of SCWO Applications

The application of the SCWO process to a great variety of materials was briefly discussed by Brunner [2] and here only a few kinds of application are to be described.

Municipal sewage excess sludge, containing 3.5 mass% solids (microorganisms and inorganic materials) and $9.72 \, \mathrm{g \, dm^{-3}}$ TOC, was treated by

Goto et al. [43] in an SCWO batch reactor. At $\geq 500°C$ and just over 100% of stoichiometric hydrogen peroxide oxidant, complete removal of TOC took place in ≤ 1 min. The resultant solid was odorless and pale red, while the liquid was odorless, transparent, and colorless. At 30 MPa and at the tested highest temperature of 600°C, the nitrogen content, which as an intermediate formed ammonium ions, was also completely turned into nitrogen gas.

The treatment of aqueous industrial wastes is a major field of application, and of the many papers dealing with it, only a few are to be discussed here. The destruction of aqueous wastes containing biocides was described by Baur et al. [44], who reported on the SUWOX process developed in Karlsruhe, Germany. A high pressure (70 MPa) and a relatively low temperature (480–500°C) ensure a high density of SCW, at which salts (e.g., > 200 g dm^{-3} NaCl) remain soluble and do not precipitate. High-pressure oxygen (70 MPa) must be produced in a separate facility if it is to replace the hydrogen peroxide otherwise used. A pilot plant with a throughput of 200 kg h^{-1} was constructed and 1000 h operation was demonstrated. The level of dioxin in the treated water was below detection limit.

A case study of an SCWO process for dealing with textile dye house wastewater was reported by Sögut and Akgün [45], based on previous development studies [46]. The operating conditions were 25 MPa and $\geq 450°C$ and used hydrogen peroxide as oxidant. Residence times in the tubular reactor were as short as 8–16 s for practically 100% removal of total organic carbon, initially at a 850 g m^{-3} level.

Wastewater arising from an acrylonitrile manufacturing plant was treated by an SCWO process, as discussed by Shin et al. [47]. At a residence time of 15 s at 552°C and 25 MPa, a 97% removal of the TOC from the wastewater was achieved. The range of TOC in the feed, 3–25 g dm^{-3}, had no appreciable effect on the efficiency of the process, provided that the oxygen amount, supplied as hydrogen peroxide, was at least (but no more than) stoichiometric.

A pilot plant for the SCWO processing of aqueous cutting oil waste with a throughput of 25 kg h^{-1} was described by Jimenez-Espadafor et al. [48]. Such wastes neither contain chlorine atoms nor salts; hence, the problem of scale formation is absent. Operating at 500°C, the process removed 97% of the chemical oxygen demand and at high feed concentration (COD > 40 g dm^{-3} of O_2), the regenerated energy was up to 71% of the available heat content.

An example of the SCWO treatment of a nonaqueous waste is that of transformer oil contaminated with polychlorinated biphenyls (PCBs) found in South America [49] and in Korea [50]. In the former study, 99.6% conversion of the TOC and the reduction of PCB level to below detection limit was achieved at 540°C and 24 MPa with a 350% oxygen excess, using hydrogen peroxide. In the Korean study, oil samples with 50 ppm PCB and 1.82% TOC and 4.07% COD were emulsified in water forming the feed.

Oxygen-rich air (92% O_2) was used as the oxidant in a pilot plant with a $30 \, kg \, h^{-1}$ capacity. The optimal temperature and pressure were 500°C and 26.4 MPa, respectively, 60 s was the optimal residence time, and 150% of the stoichiometric oxidant was the optimal amount. The destroying efficiency was >99.9% of the TOC and measured dioxin in the environment of the plant was 15% of the Korean allowed limit.

The treatment and remediation of soils contaminated with PCBs and polycyclic aromatic hydrocarbons (PAHs) by SCWO processes is the subject of a large number of reports in the last decade or so. Kronholm et al. [51] studied the treatment of PAH-contaminated soil from a decommissioned Swedish mine with a combined pressurized hot water extraction (PHWE) process coupled with SCWO. In the extraction step, water heated to 300°C at 30 MPa removed the PAHs from the soil and delivered it to the SCW unit operating at 425°C, where the PAHs were oxidized with hydrogen peroxide. In a subsequent paper [52], the authors achieved a considerably better destruction of the PAHs than initially reported (91%), namely, 97.0–99.9%, depending on the compound.

Another two-stage remediation process for PCB-contaminated soils and sediments proposed by Zhou et al. [53] involves extraction with supercritical carbon dioxide modified by methanol and subsequent SCWO treatment. The first stage operates at 50°C and 12 MPa with 5 mol% methanol, and the second stage operates at 500°C and 25 MPa. SCWO treatment of contaminated soils and sediments was compared by Haglund [54] with other available methods in terms of removal efficiency and costs and was found to be competitive.

5.3 USES OF SCW IN ORGANIC SYNTHESIS

There are many uses for SCW in organic synthesis, both as a reactant and as a solvent. Some of these have been recently reviewed [1, 2, 55–59]. The growing interest in SCW as a reaction medium is evident [57] in the number of citations and pages that increased from 60 citations and 12 pages in a 1986 review to 400 citations and 56 pages in a 1995 review to nearly 200 pages in a thematic issue of *Chemical Reviews* in 1999. Interest seems to have abated since then, as many fewer papers are nowadays devoted to organic synthetic reactions in SCW. The reactions that have been studied include hydrolysis, hydration, dehydration, hydrogenation, partial oxidation, polymerization, among others, examples of which are shown in Table 5.1. The kinetics of the reactions are, of course, much faster than that at subcritical temperatures because of the high temperature, low viscosity, and fast diffusion. An advantage of using SCW as a medium is the appreciable solubility of reactant gases, such as hydrogen and oxygen, in SCW leading to fast homogeneous reactions with them. Another advantage is the ready separation of the product from the reaction mixture by the rapid expansion of supercritical solutions process involving a sudden lowering of the pressure. For reactions in which the intermediate involves a volume reduction compared to the reactants, the high pressure used in SCW processes is another advantage. Where acid or base catalysis is required, SCW considerably provides larger concentrations of hydrogen and hydroxide ions than does the low-temperature water (see Section 2.5). It is to be noted that the examples shown in Table 5.1 are all at the lower range of temperatures of SCW, 380–400°C. At higher temperatures, undesirable side reactions such as thermal decomposition take place.

Although many useful reactions take place in SCW in the absence of a heterogeneous catalyst, some others benefit from its presence, as reviewed by Savage [57]. Hydrogenation/dehydrogenation over PtO_2 is an example,

TABLE 5.1   Examples of Organic Synthesis Reactions Carried Out in SCW

Reaction, Substrate/Product	T (°C), P (MPa)	Reference
Hydrolysis, poly(tetrahydrofurandiacetate)/ poly(tetrahydrofuran)	400	[85]
Hydrolysis, starch/monosugars	380, 24	[1]
Hydrolysis, ethyl acetate/acetic acid + ethanol	400, 23–30	[1]
Hydration, ε-aminocapronitrile/ε-caprolactam	380	[85]
Dehydration, 1,4-butanediol/tetrahydrofuran	400, 30	[85]
Dehydration, glucose/glyceraldehydes, pyruvaldehyde	400, 40	[1]
Partial oxidation, cyclohexane/cyclohexanone	400	[85]
Partial oxidation, methanol/formaldehyde, formic acid	500, 24.6	[58]
Partial oxidation, methane/methanol	480	[61]
Partial oxidation, phenol/phenoxyphenol, catechol	420	[2]
Partial oxidation, xylenes/phthalic acids	380, 23	[127]
Diels–Alder, diethylmaleate + cyclopentadiene	380	[58]
Beckman rearrangement, cyclohexanone oxime/ε-caprolactam		[59]
Beckman rearrangement, cyclohexanone oxime/ε-caprolactam		[128]
Heck coupling, styrene + iodobenzene/stilbene, phenol	377	[59]
Heck coupling, phenol/monoterpene alcohols	450, 40	[59]
Disproportionation, benzaldehyde/benzoic acid + benzyl alcohol	427, 25	[129]
Cyclotrimerization, acetylenes/benzene derivatives	400	[130]

leading to aromatization of cyclohexyl derivatives. A Pd catalyst was used in alkene–arene coupling reactions. Partial oxidation of methane to methanol in the presence of a catalyst was reviewed by Watanabe et al. [60], who showed that the selectivity toward methanol production is inversely correlated with the extent of conversion of methane, but that active catalysts, such as TiO_2, ZrO_2, and RuO_2 could perhaps ameliorate the situation, which at 450°C, a ratio of $O_2/CH_4 = 0.072$ in the feed led a 10% methane conversion to 20–40% methanol selectivity at best [61]. The degradation of waste polystyrene for the recovery of monomer styrene was studied by Kronholm et al. [62]. The best medium was NaOH in SCW at 400°C, where 57% of the styrene monomer could be recovered by means of thermal field flow fractionation.

The prospects for organic synthetic reactions in SCW looked very bright in 1999 [56]. It was concluded that ". . .the area of chemical synthesis in high temperature water is the one that will grow most rapidly. There are simply vast expanses of uncharted territory in this field. . .." These prospects appear not to have materialized in spite of the above-listed advantages that this "green" solvent provides, as only few papers on organic synthesis in SCW have been published in 2010.

5.4 USES IN POWDER TECHNOLOGY OF INORGANIC SUBSTANCES

Supercritical fluids, mainly supercritical carbon dioxide, have been applied widely to formulations of drugs and pharmaceuticals. However, such materials are too sensitive to thermal decomposition at the harsh conditions prevailing in SCW, so that there appear to be no reports of such applications in the latter medium. On the contrary, thermally stable inorganic materials have been dealt with in SCW with advantage, mainly for the production of micro- and nanoparticles of metal oxides and other materials. This subject was already reviewed in 1989 by Mason and Smith [63], then briefly by Hakuta et al. [64], and recently more fully by Hayashi and Hakuta [65].

In case of metal oxide formation, the reactions that take place in SCW before the expansion (in RESS processes) are hydrolysis of the aqueous metal salt (with alkali added to neutralize the acid formed) to form hydrolyzed species, formulated as the nominal hydroxide:

$$MX_n(aq) + nH_2O \rightarrow M(OH)_n + nHX\ (aq) \tag{5.4}$$

and dehydration of the $M(OH)_n$ so formed:

$$M(OH)_n \rightarrow MO_{n/2} + (n/2)H_2O \tag{5.5}$$

The oxide has some solubility in SCW, depending on its temperature and density; for example, at 500°C and 100 MPa, the solubilities in ppm are 1.8 for Al_2O_3, 20.0 for NiO, 90 for Fe_2O_3, 2600 for SiO_2, and 8700 for GeO_2 [63]. In RESS process in SCW, the hydrolyzed soluble solutes immediately supersaturate the solvent, then nucleation occurs, and in the third stage the particles grow.

SCW process appears to be the preferred one for the production of fine metal oxide powders by hydrothermal hydrolysis of salts, and in particular for the production of homogeneously dispersed metal oxide mixtures. Batch processes have their advantage when single-phase crystals are to be produced at low alkali concentrations and long crystal growing periods. Flow processes can be employed for controlling the crystal size and habit. Examples of metal oxide particles produced in SCW are shown in Table 5.2. It is to be noted that most of the processes were carried out at the lowest temperatures in which water exists in a supercritical state and at moderate pressures. The resulting metal oxides in batch processes have regular fine morphology, large surface areas, and good crystallinity. TiO_2 and $K_2Ti_6O_{13}$ produced can be used as photocatalysts, for instance, for water decomposition [65], and $KNbO_3$ has applications in nonlinear optical devices [66]. Luminescent properties of rare earth-doped YVO_4 particles formed by SCW synthesis were reported by Zheng et al. [67]. The manganese-doped zinc silicate is used as a phosphor for plasma display panels [65]. The nanoparticles produced in flow processes, at very short residence times, result from two features of the SCW process in

TABLE 5.2   Examples of metal oxide fine particles produced in SCW

Product	Conditions	Particle Size (nm)
TiO_2	B, ≤400°C, 24 h	20–115
$KNbO_3$	B, 400°C, 24 MPa, 2–24 h	15–42
$KTaO_3$	B, 400°C, 25 MPa, 2–48 h	1000–10 000
Zn_2SiO_4:Mn	B, 400°C, 29 MPa, 0.5–1.5 h	>1000
$CoFe_2O_4$	B, 390°C, <6 h	5
$Mg_{3.5}H_2(PO_4)_3$	B, 400–450°C, 25–32 MPa, 5–120 min	20–500
ZrO_2	F, 400°C, 30 MPa, 1.8 s	7
$BaTiO_3$	F, <420°C, 30 MPa, 0.1–40 s	13–48
ZnO	F, 390°C, 30 MPa, 0.7 s	16–57
$CoFe_2O_4$	F, <402°C, 25 MPa, 11–23 s	13–23
γ-AlOOH, γ-Al_2O_3	F, 400°C, 30–40 MPa	<20
$BaZrO_3$, cubic	F, 450–485°C, 30 MPa	<100
$BaZrO_3$, tetragonal	F, 400°C, 30 MPa	<20
CeO_2, cubic	F, 400°C, 30 MPa	<50

Adapted from Ref. [65]. B: batch process; F: flow process.

addition to those in batch processes. The rapid drop in the permittivity after injection causes very high-density homogeneous nucleation, whereas the short process times suppress crystal growth and aggregation and a narrow distribution of nanosized crystals results.

The rate of nucleation of a precipitating solute (NaCl) in SCW was studied by molecular dynamics simulation by Svishchev [68]. The nuclei, consisting of 15–30 ions, initially form amorphous structures at a rate of $10^{28}\,cm^{-3}\,s^{-1}$, which eventually form nanocrystals.

The nozzle reactor used by Lester et al. [69] produced not only metal oxide particles (TiO_2, CeO_2, and ZrO_2) with small diameters (6–13 nm) and large BET surface areas (90–195 $m^2\,g^{-1}$) but also fine metal powders, for example, of silver (9 nm diameter, 60 $m^2\,g^{-1}$ BET surface). Surface impregnation of MnO, PbO, and silver metal on alumina in a batch-type SCW process was demonstrated by Otsu and Oshima [70]. More recent examples of nanoparticles produced by SCW processes are the formation of yttrium aluminum garnet (YAG) and monoclinic $BaTeMo_2O_9$ reported by Li et al. [71, 72].

However, early applications of supercritical drying of silica gels successfully used alcohol/water mixtures that were eventually heated and pressurized to above the critical point of water [73]. Hybrid materials involving hydrophilic metal oxide particles and hydrophobic polymers can be produced by simultaneous injection into the SCW reactor of streams of the aqueous metal salt and of an organic modifier, such as decanoic acid, and supercritical water. The nanoparticles are coated by the organic modifier and can then be dispersed homogeneously in the polymerization medium or may self-assemble on a suitable substrate [74].

Lee [75] reported on a Korean pilot plant for supercritical hydrothermal synthesis of fine powders with a capacity of 30 t yr^{-1}. However, Lester et al. [69] expressed reservations concerning problems with the reliability, reproducibility, and process control of the hydrothermal synthesis of the nanoparticles in flow systems. These problems could be overcome by a new type of nozzle reactor described here.

5.5 GEOTHERMAL ASPECTS OF SCW

The history of hydrothermal synthesis of minerals was reviewed by Eugster [76], who suggested that "behind every important mineral there once was a fluid." Most mineral transformations in the upper few tens of kilometers of the earth's crust take place at a low-viscosity hydrous fluid that may also contain carbon dioxide. At depths ≥ 10 km, the release of H_2O and CO_2 from hydrous and carbonate minerals is the major source of this fluid [77]. The fluid migrates to the surface because of its lower density, but on its way it is at equilibrium with the minerals over which it passes and causes mineral alterations (metasomatism) and dissolves some of them. The solubilities of the minerals in hot pressurized water and solutions of CO_2 in it are quantities important for the understanding of the processes that take place. Aqueous fluids at elevated pressures and temperatures were recently reviewed in connection with geothermal processes by Liebscher [78]. On the other hand, the long-term storage of CO_2 in deep saline formations is envisaged as one of the alternatives to reduce the release of CO_2 greenhouse gas to the atmosphere. This was simulated by means of a two-dimensional model for CO_2 trapping and transport in certain Chinese formations [79]. The amount trapped in carbonate minerals gradually increases with time.

The solubilities of inorganic solutes in SCW, their thermodynamic properties, and their transport properties are summarized in Section 4.3 and need not be repeated here. However, the solubility of quartz (silica, SiO_2) in SCW is a key quantity for geochemical purposes and is briefly discussed here. McKenzie and Helgeson [80] discussed the data of Anderson and Burnham [81, 82] and others, noting that the dissolution reaction is

$$SiO_2(c) + nH_2O \rightarrow SiO_2 \cdot nH_2O(aq) \qquad (5.6)$$

The pressures encountered in deep geological strata are much higher than those used in the applications already discussed Sections 5.1–5.4, namely, 100 MPa and above. At 100 MPa, the solubility of silica increases only mildly with increasing temperatures: 0.24 mass% at 500°C to 0.26 mass% at 600°C, but at 200 MPa the increase is from 0.41 to 0.72 mass% over this temperature interval and solubility as high as 8 mass% was measured at 900°C and 700 MPa [81]. The solubility of corundum (alumina, Al_2O_3) in SCW was also measured (being very much less than that of quartz), as were the solubilities of these two solutes in HCl, NaCl, KCl, NaOH, and KOH solutions in SCW [82]. Chlorides decrease slightly the solubility of quartz in SCW, but alkalis increase it in proportion to the alkali content by direct reaction. The reaction was written [77] as

$$Si(OH)_4(c) + NaOH (aq) \rightarrow NaSi(OH)_5 (aq) \qquad (5.7)$$

depending on the coordination number of Si(IV) being six. Similarly, alkalis increase the concentration of Al(III) according to

$$Al(OH)_3 (c) + NaOH (aq) \rightarrow NaAl(OH)_4 (aq) \qquad (5.8)$$

No chloride complexing takes place with Si(IV) or Al(III), but other metal ions complex with chloride in supercritical brines (NaCl (aq)) existing in deep strata [76]. When metamorphic reactions are written, it is often assumed that aluminum is conserved between the solid phases, due to its extremely low solubility under neutral conditions. Still, appreciable amounts of Al(III) can be transported in high-temperature hydrothermal veins leading to the occurrence of mica and feldspars [83]. The dissolution of anorthite, $CaAl_2Si_2O_8$, the calcium endmember of plagioclase feldspar in SCW + SCD [84] is an example of such geochemical–hydrothermal processes.

At very high pressures, $\geq 1.5\,GPa$, prevailing in deep strata of the upper mantle of the earth, silica and silicates become miscible with water. This was observed in a diamond anvil cell [85], the critical temperatures for miscibility at 1.5 GPa being 550°C for nepheline, 800°C for Jadeite, and 900°C for calcium bearing granite. When such a supercritical fluid rises to shallow depths (30 km below surface), it unmixes to form a hydrous silicate melt and a water-rich fluid. Such a fluid, in particular if it contains some chloride, is likely to concentrate low-field cations such as Rb(I), Ba(II), and Pb(II) and transport them. *Ab initio* and MD simulations at 727°C and 0.9–3.6 GPa were used to study the coordination of Ti(IV) with water molecules in SCW, pertinent to prealkaline and peraluminous domains in geochemical hydrous fluids [86].

The suboceanic sediments permit hydrothermal vents to be established and in one case, the Escabana Trough, the chlorinities of the sediment pore fluids were both higher in some pores and lower in others than that of the oceanic water [87], resulting from supercritical phase separation. The critical point of seawater (407°C and 29.8 MPa) is considerably higher that that of pure water, but its required critical pressure and the temperature are exceeded respectively at the sea floor of the Escabana Trough and at depths of 200 m below the sea floor.

The importance of the partial molar volumes and enthalpies of reacting species in SCW was stressed by Helgeson [88] with regard to geochemical problems. These issues are discussed in Section 4.3.2. Helgeson and his coworkers presented extensive tables of the thermodynamic quantities involved in high-temperature and high-pressure aqueous solutions of inorganic and organic solutes [28, 89–92]. Hydrothermal sulfide mineral solubilities depend on the acidity (H_2S as reagent) or otherwise (HS^- as reagent) of SCW. Under the former conditions, $(\partial \log K/\partial T)_P > 0$ and $(\partial \log K/\partial P)_T < 0$ and under the latter conditions, $(\partial \log K/\partial T)_P < 0$ and $(\partial \log K/\partial P)_T > 0$, where K is the solubility equilibrium constant. The hydrolysis of feldspar and other aluminosilicates in SCW to form quartz and $KAl(OH)_4$ has an equilibrium constant K for which $(\partial \log K/\partial P)_T > 0$, but $(\partial \log K/\partial T)_P$ reaches a maximum [88].

Although C (graphite), CO_2, and CH_4 are the main forms of carbon in metamorphic crustal rocks, metastable persistence of crude oil, consisting mainly of asphaltenes and heavy aromatic-rich hydrocarbons, in CO_2-rich aqueous fluids was postulated by Helgeson [93]. Even biomolecules may have metastable presence in nonoxidizing hydrothermal fluids at high temperatures and pressures [94]. It is necessary to monitor the chemical potentials of H_2, CO_2, NH_3, and H_2S in SCW in order to understand the reactions involving amino acids and other biomolecules.

Simoneit [95, 96], among others, studied the organic geochemistry and petroleum generation of hydrothermal systems. At water depths > 2.9 km, the confining pressures are > 29 MPa and temperatures $> 400°C$ may be encountered, so that supercritical conditions for the oceanic water seeping into sediments are achieved. Reductive environments then alter organic compounds to petroleum hydrocarbons. The fluids move the products from high-temperature areas to low-temperature ones, where the higher molecular weight material accumulates. Bassez [97] argued that in SCW, where water dimers prevail, under very high pressure and a reduced O–O distance, a nonpolar and symmetrical dimer has the lowest energy, hence it is the most stable configuration. This should promote the solubility of nonpolar molecules, and in particular could be a medium for prebiotic synthesis of amino acids, sugars, and the bases of nucleic acids from nonpolar CH_4, CO_2, H_2S, and N_2 present in deep ocean hydrothermal vents. Sudden expansion of SCW rising out of hydrothermal vents may create shock waves and the minerals in the walls of the vents may act as catalysts for the formation of primordial HCN and HCHO and thus of the building blocks of biomolecules. These postulates and experiments that could substantiate them were subsequently reviewed by Bassez [98].

5.6 APPLICATION OF SCW IN NUCLEAR REACTORS

Water plays an important role in nuclear reactor technology in a dual capacity: as a neutron moderator and as a primary or secondary coolant. Elastic collisions of the fast neutrons, resulting from the nuclear fission occurring in reactor fuel rods, with the hydrogen atoms of water molecules slow them down to thermal energies at which they efficiently propagate the fission chain reaction. Light water (H_2O) has a small but finite thermal neutron absorption cross section so that heavy water (D_2O) with a negligible cross section may replace it. In the swimming pool research reactors, the role of the water ends in the primary cooling: it extracts the heat from the fuel rods and dissipates it to a large volume of water. In power reactors, the water that is in contact with the fuel rods transfers it to a secondary loop that provides the steam to operate the turbines of the electric generators. The thermodynamic efficiency of the electric power generation from the nuclear fissions in the fuel rods depends directly on the temperature difference between the fluid in the primary cooling circle and the outlet vapors from the turbines. Hence, there has been a tendency over the years to move to boiling water reactors (BWRs), then to pressurized water reactors (PWRs) operating above the normal boiling point of water, and finally to supercritical water reactors (SCWR) or SWR) (Fig. 5.3) in a "generation IV" nuclear reactor concept. The expected thermal efficiency of such reactors/power plants is 45–50% [99]. Water is not necessarily the best coolant and neutron moderator when very high temperatures are aimed at: carbon dioxide and molten sodium have been proposed, in particular when no moderator is wanted, in fast (neutron) breeders. Then the fluid that extracts the heat from the reactor core may still transfer it advantageously to SCW in the secondary loop.

Wright et al. [100, 101] discussed in the mid-1960s the state-of-the-art at the time of SCW as a reactor coolant. The subject has since been taken up to a large extent by Chinese and Japanese researchers. For example, Oka et al. [102] suggested a design of a direct cycle light water reactor operated at supercritical conditions. The design aimed at 1.145 GW electrical output at a thermal efficiency of 41.6%, much larger than BWR or PWR reactors. The operating pressure was 25 MPa and the inlet and outlet water temperatures were 310°C and 416°C, respectively, while in the interior of the pressure vessel the temperature was 385°C, near the maximum of the heat capacity of water, so that heat removal from the fuel rods (5.6% enriched UO_2) was highly efficient. The decrease of density of the water between the inlet and outlet of the pressure vessel was from 725 to 137 kg m^{-3}. In a subsequent paper, Oka et al. [103] proposed a similar design of a fast breeder reactor with a 1.246 GW electrical output and the same high thermal efficiency, that is, 41.6%. The outlet SCW temperature was 431°C and its density decreased to

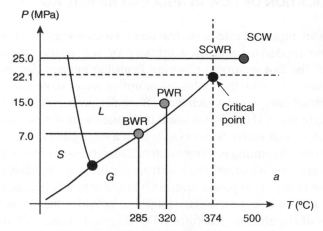

FIGURE 5.3 Schematics of the temperatures and pressures used in light water moderated and cooled nuclear reactors. Adapted with permission from Ref. [125].

121.5 kg m^{-3}, with the same inlet water states as in the previous design. The advantage in the use of SCW was the operation with a single water (supercritical) phase in the reactor core and turbine, simplifying the design considerably, and the large thermal efficiency that could be achieved.

Zhao et al. in Refs [104, 105] dealt with the stability of the U.S. reference SCW cooled nuclear reactors. One concern was with the drastic change in density of the fluid in the reactor core, in particular at the start-up stage, where a two-phase steam + liquid water system rather than a single-phase SCW occurs. A European design of a high-performance light water reactor (HPLWR) was reported by Schulenberg et al. [106] involved a three-pass SCW cooled and moderated core with a target output of 1 GW electric power, operating at 25 MPa with input water at 280°C and output water at 500°C as live steam. A main concern of the design was to stay within the maximal temperature of 620°C allowed for the fuel cladding. A more recent follow-up of this HPLWR concept was published by Maraczy et al. [107], taking into account coupled neutronic–hydraulic analysis. A recent Canadian concept uses uranium carbide pellets as the fuel, sheathed in Inconel-600, with channel inlet and outlet temperatures of 350°C and 625°C, respectively, operating with SCW at 25 MPa [99].

Many concepts and designs of SCWRs have been reported, as reviewed, for example, by Corradini [108]. An item that was stressed as requiring investigation was the radiolysis of the water in the reactor core and the effect of the radiolytic products on the construction materials [109, 110]. The addition of hydrogen for the neutralization of the primary radiolytic species, found useful for lower temperature PWRs, may not be adequate for the important temperature range of 350–450°C in SCWRs. A major concern with all SCW-based nuclear reactor designs is the corrosion of the fuel rod cladding and other construction materials by the SCW [109, 111]. This subject of corrosion is discussed in Section 5.7. In spite of extensive current efforts, as of 2010 "supercritical water (SCW) has never been used in nuclear power applications" [112] and only designs have been reported.

5.7 CORROSION PROBLEMS WITH SCW

Supercritical water is a highly reactive substance, in particular if it contains some dissolved oxygen. Its effect on organic materials – supercritical water oxidation for the complete destruction of noxious and hazardous substances – has already been described in Section 5.2. However, the common construction materials of vessels and piping in which SCW is handled are also subject to attack by this reactive medium, and many foreseen applications have been hampered because of this. For example, quartz or silica glass that is inert to many reactive media, such as concentrated mineral acids (but not HF!), has an appreciable solubility in SCW (Section 5.5). Ordinary stainless steel is vulnerable to SCW if it contains chlorides, but special alloys and ceramics may be stabilized against attack. This subject has, therefore, drawn much attention by researchers who plan to apply SCW in any technology and an extensive literature on this subject exists that is only briefly considered here. Recent reviews of corrosion problems in SCW are those of Kritzer [113] and of Ropital [114].

Typical construction materials for applications of SCW are stainless steels, nickel-based alloys, titanium, tantalum, and ceramics. The factors controlling their corrosion were studied by Kritzer et al. [115]. The stability of the protecting oxide layers depends on the ionic concentrations in the SCW due to dissolved salts, acids, and bases (Section 4.3), on the solubility of gases in SCW (Section 4.1), and on the solubility of the corrosion products in SCW. Although higher temperatures are generally conducive to corrosion, in case of SCW such temperatures are related (at a given pressure) to lower densities and severe corrosion takes place only at densities $\geq 300 \, kg \, m^{-3}$.

In the absence of sulfate and phosphate, the corrosion resistance of titanium to SCW $+$ HCl is the best even up to 600°C. Nickel-based alloys are resistant to SCW at densities $\leq 300 \, kg \, m^{-3}$, and hence are suitable for SCWO applications. Some alumina and (CeO_2-stabilized) zirconia ceramics are resistant to corrosion in SCW but not in the presence of sulfuric acid. In alkaline solutions, the corrosion resistance of nickel-based alloys is several orders of magnitude larger than in acidic solutions, even in the presence of chloride [115]. A long time (800 h) test of nickel-based alloy 625 (62% Ni, 22% Cr, 9% Mo, 3.5%Nb, 1.5% Fe, all mass%) under SCWO conditions (1.44 mol kg^{-1} O_2 in SCW) showed very low corrosion up to 420°C and tolerable corrosion (0.5 mm yr^{-1}) at 420–500°C [116]. NiO protective layers for both nickel-based alloys and stainless steels are resistant to corrosion in high-density SCW at ≥ 450°C in the presence of oxygen and acids (it is the Cr(III) oxidizing to Cr(VI) that is responsible for corrosion). Phosphate, fluoride, and hydroxide anions cause some passivation at 350–450°C [117]. High-purity alumina (99.9 mass%) and zirconia-toughened alumina successfully resist corrosion in SCWO environments. Sulfuric acid, compared to hydrochloric and phosphoric acids, is the most aggressive medium toward these materials [118]. More recently, niobium was suggested as a corrosion-resistant material under SCWO conditions [119] and a double-walled reactor, with titanium exposed to the corrosive SCWO that is strengthened against pressure by an outer Hastelloy C-274 sheath, was also demonstrated [120].

Although these studies are mainly concerned with the corrosion of construction materials in SCWO processes, Marrone and Hong [121] dealt with both these and SCW gasification processes (Section 5.1.1). In the former processes, strong oxidizing media prevail, whereas in the latter ones hydrogen and a reducing medium are formed. These authors discussed the corrosion control methods suitable for such applications. They, among others [113], pointed out that the corrosion problems are more severe at subcritical temperatures than in SCW because the higher density, permittivity, and ionic dissociation of the water in the former medium permit larger concentrations of well-hydrated aggressive ions to exist. Therefore, the corrosion problems in preheater and cooled down piping and devices would be more severe than in the reactor itself. The approaches suggested [121] for corrosion control in SCWO processes include the prevention of corrosive species from reaching solid surfaces, for example, by using a thin water film running down the reactor wall (transpiring wall cooled reactor) or absorbing such species on solid particles in a fluidized bed reactor. The choice of the construction materials is, of course, crucial and corrosion-resistant liners and coatings are possible solutions. Preneutralization and dilution of wastes that would produce aggressive acids should also prevent corrosion. Similar strategies are generally applicable also to SCWG processes, although the materials of the reactor and piping would be different. Thus, although titanium is well suited for SCWO reactors in the presence of HCl, it is subject to hydrogen embrittlement in SCWG processes. Graphite and silicon carbide linings that have good heat transfer properties may be useful for SCWG reactors, and graphite was selected by Richard and Poirier [122] as a very good material, with alumina and silicon carbide being strongly corroded by wet waste.

Corrosion problems in a third type of application, SCW moderated and cooled nuclear reactors of generation IV (Section 5.6), have also received much attention, as reviewed by Cabet et al. [123] and by Sun et al. [124]. The presence of corrosive solutes, such as oxygen or acids, in SCW is avoided in this application, but even very pure SCW has corrosive properties in spite of its low permittivity and ionic dissociation. Additional points to consider are the very thin fuel element claddings and the irradiation products of the water, the radiolysis of which produces oxidizing species such as O_2 and $OH^\bullet$ radicals [125]. Austenitic stainless steel has been proposed by Nakazomo et al. as a cladding material for the nuclear fuel, since it resists corrosion under irradiated SCW conditions up to 600°C [112]. Austenitic stainless steel and nickel-based alloy 625 were also considered as suitable by Penttila et al. [111]. Alumina, applied as a thin coating on an iron-based alloy (P91) by spray pyrolysis, proved to be an effective corrosion protection. Zirconia deposited by plasma electrolytic oxidation on a zircaloy cladding also protected the latter to some extent in SCW, according to Hui et al. [126].

Corrosion problems and methods of minimizing and controlling them are seen to be of great importance in all applications of SCW. It has been suggested that collaboration between geochemists (dealing with solubilities of inorganic substances in SCW), corrosion scientists, and potential users of SCW should be fruitful for better understanding of the dissolution behavior and for finding solutions to problems of corrosion of construction materials of devices operation in pressurized high-temperature water, that is, SCW [113].

5.8 SUMMARY

Applications of SCW are based on several aspects of SCW: its being a "green" solvent, its properties being tunable by changes of the temperature and pressure, and its solvent power for a variety of substances. The general conditions of the application of SCW are $380 \leq t/°C \leq 700°C$ and $25 \leq P/MPa \leq 40$, although higher temperatures and pressures may also be used. Once a reaction or separation has been effected in SCW, the general technique for obtaining the desired product or for the disposal of unwanted waste is RESS. RESS, with the concomitant cooling, precipitates the solutes if solid or produces a separate layer if liquid that can be handled and removed. The energy balance and the corrosive properties of SCW toward materials of construction are major aspects of concern.

Conversion of biomass, consisting mainly of cellulose, hemicellulose, lignin, and possibly water (wet biomass) to biofuel is a cardinal use of SCW. In an SCW gasification of biomass process, the gas produced is generally a mixture of hydrogen, methane, carbon monoxide, and carbon dioxide, called syn-gas. Wet biomass with a large water content (≤ 80 mass%) can be processed. The carbon dioxide produced dissolves readily in the SCW, letting the hydrogen and methane to be collected as gases. The water gas shift reaction (5.1) may be invoked to get rid of the carbon monoxide and produce more hydrogen. At the lower temperature range of SCW, near 400°C, methane is the main gaseous product, but near 650°C hydrogen becomes the major combustible product. A catalyst is useful for the reduction of the production of undesirable tar and coke by-products. A demonstration plant, VERENA, in Germany, is schematically shown in Fig. 5.1. The process was modeled with 13 relevant reactive species: H_2O, H_2, CO, CO_2, CH_4, O_2, NH_3, H_2S, CH_3CHO, CH_3CO_2H, C_6H_5OH, $C(s)$, and INO (nonreacting). Applications of the SCWG process, dealing with industrial waste streams and with sewage, have also been presented.

The lignin, a hydroxyaromatic polymer, content of plant material biomass prevents efficient enzymatic attack of the cellulose that cannot be readily fermented to ethanol. SCW pretreatment, catalyzed by carbon dioxide, acting as an acid, hydrolyzes the polymeric hydroxyaromatic substances to a fermentable broth, but mild SCW conditions of 400°C and 25 MPa need to be used, in order to avoid further degradation of the water-soluble oligosaccharides to be fermented in a second stage. Extraction of bitumen from low-lying oil sand layers is carried out with the steam-assisted gravity drainage process, so that preheating and mixing with water have already been effected for subsequent processing by SCW. The latter constitutes not only a solvent but also a reactant that supplies the hydrogen atoms needed to cap radical intermediates, thus preventing coke production. Studies on such processes have been undertaken mainly by Chinese and Japanese groups.

Supercritical water oxidation aiming to destroy hazardous and noxious organic substances and convert them to environment-friendly products is a principal application of SCW. The good solubility of organic substances and oxygen in SCW at the typical operation conditions of 650°C and 25 MPa lead to the oxidation reaction occurring in a homogeneous phase without problems of mass transfer and interface crossing. The heat generated in the exothermic oxidation process is able to sustain all of its energy requirements and even leave some for external use. Several industrial size SCWO plants as well as demonstration and pilot plants have by now been operative. The SCWO process consists of four main steps: feed preparation, reaction, salt separation, and heat recovery, as shown in the schematic flow sheet in Fig. 5.2. The feed is characterized by its total organic carbon content or by its chemical oxygen demand for total oxidation. The feed preparation includes the choice of the oxidant: oxygen as air, liquid oxygen, or hydrogen peroxide, or even produced *in situ* by electrolysis, the efficiency of the process not depending on this choice. There is a play-off between the temperature of the process and the residence time: complete oxidation with stoichiometric amounts of the oxidant occurs at 650°C at a residence time < 50 s. The salts produced from halogen, sulfur, phosphorus, and in some cases nitrogen atoms in the organic feed are kept in solution in the SCW if its density is sufficiently high, since if they precipitate out, they may plug the piping. Remedies to such situations were suggested, for example, use of KOH rather than NaOH as the base used for the neutralization of acids produced in the SCWO process. The energy balance is calculated for an organic feed consisting of $C_cH_hO_o$ by Eq. (5.3). The application of the SCWO process to a great variety of materials has been discussed in Section 5.2.2.

There are many uses for SCW in organic synthesis, both as a reactant and as a solvent. The reactions that have been studied include hydrolysis, hydration, dehydration, hydrogenation, partial oxidation, and polymerization, among others, and examples are shown in Table 5.1. However, the prognosis of 1999 that "...the area of chemical synthesis in high temperature water is the one that will grow most rapidly. There are simply vast expanses of uncharted territory in this field..." appears not to have materialized so far.

Thermally stable inorganic materials have been dealt with in SCW for the production of micro- and nanoparticles. The reactions that take place in SCW in the case of metal oxide formation before the RESS are hydrolysis of the aqueous metal salt (5.4) and dehydration of the $M(OH)_n$ so formed (5.5). When the single-phase crystals are to be produced, batch processes are used at low alkali concentrations and long crystal growing periods. Flow processes can be employed for controlling the crystal size and habit. Examples of metal oxide particles produced in SCW are shown in Table 5.2.

The relevance of SCW to geochemical processes is evident in the fact that most mineral transformations in the upper few tens of kilometers of the earth's crust take place in a hydrous fluid that may also contain carbon dioxide, due to the release of H_2O and CO_2 from hydrous and carbonate minerals. As the fluid migrates to the surface, having a lower density, it causes mineral alterations and dissolves some of them. The solubilities of the minerals and solutions of CO_2 in SCW are needed for the understanding of these processes. The solubility of silica in SCW is a key quantity for geochemical purposes, the dissolution reaction being as (5.6). Since the pressures encountered in deep geological strata are much higher than those used in the applications of SCW, appreciable solubilities are attained. At 600°C and 100 MPa, the solubility of silica is 0.26 mass% and at 200 MPa it is 0.72 mass% reaching 8 mass% at 900°C and 700 MPa. The solubility of alumina in SCW is very much less than that of silica, hence its possible use as a protective layer in SCW reactors. Chlorides decrease slightly the solubility of silica in SCW, but alkalis increase it in proportion to the alkali content by direct reaction. Helgeson with coworkers presented extensive tables of the thermodynamic quantities involved in high-temperature and high-pressure aqueous solutions of inorganic and organic solutes pertaining to geochemical conditions.

Crude oil, consisting mainly of asphaltenes and heavy aromatic-rich hydrocarbons, may persist in CO_2-rich aqueous fluids and even biomolecules may have metastable presence in nonoxidizing hydrothermal fluids. For oceanic water seeping into sediments at depths > 2.9 km, the confining pressures and temperatures correspond to supercritical conditions. Reductive environments then alter organic compounds to petroleum hydrocarbons and could also be a medium for prebiotic synthesis of amino acids, sugars, and the bases of nucleic acids.

The thermodynamic efficiency of the electric power generation from the nuclear fission reactors depends directly on the temperature of the fluid in the primary cooling circle. Reactor designs moved from boiling water reactors to pressurized water reactors, and in a "generation IV" nuclear reactor concept to supercritical water reactors. The advantage of the use of SCW was the operation with a single water (supercritical) phase in the reactor core and turbine, simplifying the design considerably, and the large thermal efficiency, $> 40\%$, that could be achieved. Recent designs consider water inlet and outlet temperatures of 350°C and 625°C, respectively, operating with SCW at

25 MPa. Concerns in the designs were the drastic change in the density of the fluid, from $>700\,\mathrm{kg\,m^{-3}}$ in the core to $<150\,\mathrm{kg\,m^{-3}}$ at the outlet, and the start-up stage, where a two-phase steam $+$ liquid water system rather than a single-phase SCW occurs. Another major concern was corrosion problems, associated not only with the general aggressiveness of SCW but also with its radiolytic products. So far generation IV power plants have not emerged from the conceptual and design stages.

The corrosion problems arising in applications of SCW depend on whether a reductive situation occurs, as in SCWG processes where hydrogen is a desired product, or an oxidative one in SCWO processes. Construction materials therefore may differ between such applications. Many metal alloys have been tested with regard to their withstanding corrosion in SCW. Typical construction materials for applications of SCW are stainless steels, nickel-based alloys, titanium, and ceramics. The stability of the protecting oxide layers in SCW depends on the ionic concentrations and on the solubility of gases and the corrosion products in it. Severe corrosion in SCW may take place only at densities $\geq 300\,\mathrm{kg\,m^{-3}}$, but less at lower densities even at higher temperatures. Thus, the corrosion problems are more severe at subcritical temperatures than in SCW; therefore, the corrosion problems in preheater and cool-down piping and devices would be more severe than in the reactor itself. Nickel-based alloy 625 and alumina-coated steel are suitable materials for SCWO reactors. Corrosion control in SCWO processes includes the prevention of corrosive species from reaching solid surfaces in a transpiring wall cooled reactor by using a thin water film running down the reactor wall. Collaboration between geochemists (dealing with solubilities of inorganic substances in SCW), corrosion scientists, and potential users of SCW should be fruitful for finding solutions to corrosion problems of construction materials of devices using SCW.

REFERENCES

1. G. Brunner, *J. Supercrit. Fluids* **47**, 373 (2009).

2. G. Brunner, *J. Supercrit. Fluids* **47**, 382 (2009).

3. A. Loppinet-Serani, C. Aymonier, and F. Cancell, *J. Chem. Technol. Biotechnol.* **85**, 583 (2010).

4. M. J. Cocero, E. Alonso, M. T. Sanz, and F. Fdz-Polanco, *J. Supercrit. Fluids* **24**, 37 (2002).

5. K. Arai, R. L. Smith, Jr., and T. M. Aida, *J. Supercrit. Fluids* **47**, 628 (2009).

6. H. Schmieder, J. Alben, N. Boukis, E. Dinjus, A. Kruse, M. Kluth, G. Petrich, E. Sandri, and M. Schacht, *J. Supercrit. Fluids* **17**, 145 (2000).

7. D. C. Elliot, *Biofuels Bioprod. Biorefin.* **2**, 254 (2008).

8. A. Kruse, *Biofuels Bioprod. Biorefin.* **2**, 415 (2008).

9. A. Kruse, *J. Supercrit. Fluids* **47**, 391 (2009).

10. E. Gasafi, L. Meyer, and L. Schebek, *J. Ind. Ecol.* **7**, 75 (2003).

11. T. Sato, S. Kurosawa, R. L. Smith, Jr., T. Adschiri, and K. Arai, *J. Supercrit. Fluids* **29**, 113 (2004).

12. H. Tang and K. Kitagawa, *Chem. Eng. J.* **106**, 261 (2005).

13. S. Letellier, F. Marias, P. Cezac, and J. P. Serin, *J. Supercrit. Fluids* **51**, 382 (2010).

14. I.-G. Lee, M.-S. Kim, S.-K. Ihm, *Ind. Eng. Chem. Res.* **41**, 1182 (2002).

15. W. Feng, H. J. van der Kooi, and J. de S. Arons, *Chem. Eng. J.* **98**, 105 (2004).

16. M. B. Garcia Jarana, J. Sanchez-Oneto, J. R. Portela, E. Nebot Sanz, E. J. M. de la Ossa, *J. Supercrit. Fluids* **46**, 329 (2008).

17. X. Xu, Y. Matsumura, J. Steinberg, and M. J. Antal, Jr., *Ind. Eng. Chem. Res.* **35**, 2522 (1996).

18. L. J. Guo, Y. J. Lu, X. M. Zhang, C. M. Ji, Y. Guan, and A. X. Pei, *Catal. Today* **129**, 275 (2007).

19. B. Yan, J. Wu, C. Xie, F. He, and C. Wei, *J. Supercrit. Fluids* **50**, 155 (2009).

20. E. Kipçak, O. Ö. Sögüt, and M. Akgün, *Proceedings of the 9th Conference on Supercritical Fluids*, Sorrento, 2010, p. 413.

21. O. Ö. Sögüt, E. Kipçak, T. Arslan, and M. Akgün, *Proceedings of the 9th Conference on Supercritical Fluids*, Sorrento, 2010, p. 427.

22. T. Nakata, H. Miyafuji, and S. Saka, *Appl. Biochem. Biotechnol.* **129–132**, 476 (2006).

23. C. Schacht, C. Zetzl, and G. Brunner, *J. Supercrit. Fluids* **46**, 300 (2008).

24. K. Ehara and S. Saka, *Cellulose* **9**, 301 (2002).

25. H. Miyafuji, T. Nakata, K. Ehara, and S. Saka, *Appl. Biochem. Biotechnol.* **121–124**, 963 (2005).

26. K. Yoshida, J. Kusali, K. Ehara, and S. Saka, *Appl. Biochem. Biotechnol.* **121–124**, 795 (2005).

27. M. Sasaki, A. Adschiri, and K. Arai, *AIChE J.* **50**, 192 (2004).

28. J. P. Amend and H. C. Helgeson, *Geochim. Cosmochim. Acta* **61**, 11 (1997).

29. M. Watanabe, H. Hirakoso, S. Sawamoto, A. Adschiri, and K. Arai Wei, *J. Supercrit. Fluids* **13**, 247 (1998).

30. H. Hu, J. Zhang, S. Guo, and G. Chen, *Fuel* **78**, 645 (1999).

31. L.-Q. Zhao, Z.-M. Cheng, Y. Ding, P.-Q. Yuan, S.-X. Lu, and W.-K. Yuan, *Energy Fuels* **20** 2067 (2006).

32. Z.-M. Cheng, Y. Ding, L.-Q. Zhao, P.-Q. Yuan, and W.-K. Yuan, *Energy Fuels* **23**, 3178 (2009).

33. M. Morimoto, Y. Sugimoto, Y. Saotome, S. Sato, and T. Takanohashi, *J. Supercrit. Fluids* **55**, 223 (2010).

34. M. Sasaki, Wahyudiono, T. Shiraishi, and M. Goto, *Proceedings of the 9th Conference on Supercritical Fluids*, Sorrento, 2010, p. 333.

35. T. Sato, S. Mori, M. Watanabe, M. Sasaki, and N. Itoh, *J. Supercrit. Fluids* **55**, 232 (2010).

36. M. D. Bermejo, D. Rincon, V. Vazquez, and M. J. Cocero, *CI&CEQ* **13**, 79 (2007).

37. B. Veriansyah and J.-D. Kim, *J. Environ. Sci.* **19**, 513 (2007).

38. B. Misch, A. Firus, and G. Brunner, *J. Supercrit. Fluids* **17**, 227 (2000).

39. B. D. Phenix, J. L. DiNero, J. W. Tester, J. B. Howard, and K. A. Smith, *Ind. Eng. Chem. Res.* **41**, 624 (2002).

40. M. Hodes, P. A. Marrone, G. T. Hong, K. A. Smith, and J. W. Tester, *J. Supercrit. Fluids* **29**, 265 (2004).

41. P. A. Marrone, M. Hodes, K. A. Smith, and J. W. Tester, *J. Supercrit. Fluids* **29**, 289 (2004).

42. S.-I. Kawasaki, T. Oe, S. Itoh, A. Suzuki, K. Sue, and K. Arai, *J. Supercrit. Fluids* **42**, 241 (2007).

43. M. Goto, T. Nada, A. Ogata, A. Kodama, and T. Hirose, *J. Supercrit. Fluids* **13**, 277 (1998).

44. S. Baur, H. Schmidt, A. Krämer, and J. Gerber, *J. Supercrit. Fluids* **33**, 149 (2005).

45. O. Ő. Sögut and M. Akgün, *J. Chem. Technol. Biotechnol.* **85**, 640 (1947).

46. O. Ő. Sögut and M. Akgün, *J. Supercrit. Fluids* **43**, 106 (2007).

47. Y. H. Shin, N. C. Shin, B. Veriansyah, J. Kim, and Y.-W. Lee, *J. Hazard. Mater.* **163**, 1142 (2009).

48. F. Jimenez-Espadafor, J. R. Portela, V. Vadillo, J. Sanchez-Oneto, J. A. B. Villanueva, M. T. Garcia, and E. J. M. de la Ossa, *Ind. Eng. Chem. Res.* **50**, 775 (2011).

49. V. Marulanda and G. Bolanos, *J. Supercrit. Fluids* **54**, 258 (2010).

50. K. Kim, S. H. Son, K. Kim, K. Kim, Y.-C. Kim, *J. Supercrit. Fluids* **54**, 165 (2010).

51. J. Kronholm, J. Kalpala, K. Hartonen, and M.-J. Riekkola, *J. Supercrit. Fluids* **23**, 123 (2002).

52. J. Kronholm, T. Kuosmanen, K. Hartonen, and M.-J. Riekkola, *Waste Manag.* **23**, 253 (2003).

53. W. Zhou, G. Anitescu, P. A. Rice, and L. L. Tavlarides, *Environ. Prog.* **23**, 222 (2004).

54. P. Haglund, *Ambio* **36**, 467 (2007).

55. H. Bureau and H. Keppler, *Earth Plan. Sci. Lett.* **165**, 187 (1999).

56. P. E. Savage, *Chem. Rev.* **99**, 603 (1999).

57. P. E. Savage, *Catalysis Today* **62**, 167 (2000).

58. A. A. Galkin and V. V. Lunin, *Russ. Chem. Rev.* **74**, 21 (2005).

59. M. Sato, Y. Ikushima, K. Hatakeda, and R. Zhang, *Anal. Sci.* **22**, 1409 (2006).

60. M. Watanabe, T. Sato, H. Inomata, R. L. Smith, Jr., K. Arai, A. Kruse, and E. Dinjus, *Chem. Rev.* **104**, 5803 (2004).

61. C. N. Dixon and M. A. Abraham, *J. Supercrit. Fluids* **5**, 269 (1992).

62. J. Kronholm, P. Vastamäki, R. Räsänen, A. Ahonen, K. Hartonen, and M.-L. Riekkola, *Ind. Eng. Chem. Res.* **45**, 3029 (2006).

63. D. W. Mason and R. D. Smith, *J. Am. Ceram. Soc.* **72**, 871 (1989).

64. Y. Hakuta, H. Hayashi, and K. Arai, *Curr. Opin. Solid State Mater. Sci.* **7**, 341 (2003).

65. H. Hayashi and H. Hakuta, *Materials* **3**, 3794 (2010).

66. B. Li, Y. Hakuta, and H. Hayashi, *J. Supercrit. Fluids* **33**, 254 (2005).

67. Q. X. Zheng, B. Li, M. Xue, H. D. Zhang, Y. J. Zhan, W. S. Pang, X. T. Tao, and M. H. Jiang, *J. Supercrit. Fluids* **46**, 123 (2008).

68. I. M. Svishchev, *Proceedings of the 9th Conference on Supercrit. Fluids*, Sorrento, 2010, p. 421.

69. E. D. Lester, P. Blood, J. Denyher, D. Giddings, B. Azzopardi, and M. Poliakoff, *J. Supercrit. Fluids* **37**, 209 (2006).

70. J. Otsu and Y. Oshima, *J. Supercrit. Fluids* **33**, 61 (2005).

71. B. Li, Q. M. Zheng, and X. T. Tao, *Proceedings of the 9th Conference on Supercritical Fluids*, Sorrento, 2010, p. 445.

72. Q. X. Zheng, B. Li, and X. T. Tao, *Proceedings of the 9th Conference on Supercritical Fluids*, Sorrento, 2010, p. 345.

73. R. A. Landise and D. W, Johnson, J., *J. Non-cryst. Solids* **79**, 155 (1986).

74. T. Adschiri, S. Takami, K. Minami, K. Ichikawa, T. Yamagata, K. Miyata, T. Morishita, M. Ueno, T. Okada, and H. Oshima, *Proceedings of the 9th Conference on Supercritical Fluids*, Sorrento, 2010, p. 433.

75. Y.-W. Lee, *Proceedings of ISHR & ICSTR*, Sendai, Japan, 2006.

76. H. P. Eugster, *Am. Miner.* **71**, 655 (1986).

77. J. V. Walther, *Pure Appl. Chem* **58**, 1585 (1986).

78. A. Liebscher, *Geofluids* **10**, 3 (2010).

79. W. Zhang, Y. Li, T. Xu, H. Cheng, Y. Zheng, and P. Xiong, *Int. J. Greenhouse Gas Control* **3**, 161 (2009).

80. W. F. McKenzie and H. C. Helgeson, *Geochim. Cosmochim. Acta* **48**, 2167 (1984).

81. G. M. Anderson and C. W. Burnham, *Am. J. Sci.* **263**, 494 (1965).

82. G. M. Anderson and C. W. Burnham, *Am. J. Sci.* **265**, 12 (1967).

83. A. B. Woodland and J. V. Walther, *Geochim. Cosmochim. Acta* **51**, 365 (1987).

84. M. Sorai and M. Sasaki, *Am. Mineral.* **95**, 853 (2010).

85. D. Bröll, C. Kaul, A. Krämer, P. Krammer, T. Richter, M. Jung, H. Vogel, and P. Zehner, *Angew. Chem., Int. Ed.* **38**, 2998 (1999).

86. J. van Sijl, N. L. Allan, G. R. Davies, and W. van Westrenen, *Geochim. Cosmochim. Acta* **74**, 2797 (2010).

87. K. L. Von Damm, C. M. Parker, A. A. Zierenberg, M. D. Lilley, E. J. Olson, D. A. Clague, and J. S. McClain, *Geochim. Cosmochim. Acta* **69**, 4971 (2005).

88. H. C. Helgeson, *Geochim. Cosmochim. Acta* **56**, 3192 (1992).

89. E. L. Shock and H. C. Helgeson, *Geochim. Cosmochim. Acta* **54**, 915 (1990).

90. D. A. Sverjensky, E. L. Shock, and H. C. Helgeson, *Geochim. Cosmochim. Acta* **61**, 1359 (1997).

91. H. C. Helgeson, C. E. Owens, A. M. Knox, and R. L. Laurent, *Geochim. Cosmochim. Acta* **62**, 985 (1998).

92. D. E. LaRowe and H. C. Helgeson, *Geochim. Cosmochim. Acta* **70**, 4680 (2006).

93. H. C. Helgeson, *Can. Mineral.* **29**, 707 (1991).

94. H. C. Helgeson and J. P. Amend, *Thermochim. Acta* **245**, 89 (1994).

95. B. R. T. Simoneit, *Geochim. Cosmochim. Acta* **57**, 3231 (1993).

96. B. R. T. Simoneit, in R. Ikan, Ed., *Natural and Laboratory Simulated Thermal Geochemical Processes*, Kluwer, Dordrecht, 2003, p. 1.

97. M.-P. Bassez, *J. Phys. Condens. Mater.* **15**, L353 (2003).

98. M.-P. Bassez, *Comput. Rend. Chim.* **12**, 801 (2009).

99. L. Grande, B. Villamere, L. Allison, S. Mikhael, A. Rodriguez-Prado, and I. Pioro, *J. Eng. Gas Turb. Power* **133**, 022901 (2011).

100. J. H. Wright and J. C. Gray, *Proc. Am. Power Conf.* **27**, 221 (1965).

101. J. H. Wright, and J. F. Patterson, *Proc. Am. Power Conf.* **28**, 139 (1967).

102. Y. Oka, S. Koshizuka, and T. Yamasaki, *Nucl. Sci. Technol.* **29**, 585 (1992).

103. Y. Oka, T. Jevremovic, and S. Koshizuka, *Nucl. Sci. Technol.* **31**, 83 (1994).

104. J. Zhao, P. Saha, and M. S. Kazimi, *Nucl. Technol.* **158**, 158 (2007).

105. J. Zhao, P. Saha, and M. S. Kazimi, *Nucl. Technol.* **164**, 20 (2008).

106. T. Schulenberg, J. Starflinger, and J. Heinecke, *Prog. Nucl. Energy* **50**, 526 (2008).

107. C. S. Maraczy, G. Y. Hegyi, G. Hordosy, and E. Temesvary, *Prog. Nucl. Energy* **53**, 267 (2011).

108. M. L. Corradini, *Nucl. Technol.* **167**, 145 (2009).

109. D. Guzonas, P. Tremaine, and J.-P. Jay-Gerin, *Power Plant Chem.* **11**, 284 (2009).

110. J. A. Wilson, R. Pathania, and S. Hettiarachchi *J. Nucl. Mater.* **392**, 230 (2009).

111. S. Penttila, A. Toivonen, L. Heikinheimo, and R. Novotny, *Nucl. Technol.* **170**, 261 (2010).

112. Y. Nakazomo, T. Iwai, and H. Abe, *J. Phys. Conf. Ser.* **215**, 012094 (2010).

113. P. Kritzer, *J. Supercrit. Fluids* **29**, 1 (2004).

114. F. Ropital, *Mater. Corros.* **60**, 495 (2009).

115. P. Kritzer, N. Boukis, and E. Dinjus, *J. Supercrit. Fluids* **15**, 205 (1999).

116. P. Kritzer, N. Boukis, and E. Dinjus, *J. Mater. Sci. Lett.* **18**, 1845 (1999).

117. P. Kritzer, N. Boukis, and E. Dinjus, *Corrosion* **56** 1093 (2000).

118. M. Schacht, N. Boukis, and E. Dinjus, *J. Mater. Sci.* **35**, 6251 (2000).

119. E. Asselin, A. Alfantazi, and S. Rogak, *Corros. Sci.* **52**, 118 (2009).

120. B. Veriansyah, J.-D. Kim, and J.-C. Jong, *J. Ind. Eng. Chem.* **15**, 153 (2009).

121. P. A. Marrone and G. T. Hong, *J. Supercrit. Fluids* **51**, 83 (2009).

122. T. Richard and J. Poirier, *Adv. Sci. Technol.* **72**, 129 (2010).

123. C. Cabet, J. Jang, J. Konys, and P. F. Tortorelli, *MRS Bull.* **34**, 35 (2009).

124. C. Sun, R. Hui, W. Qu, and S. Yick, *Corros. Sci.* **51**, 2508 (2009).

125. G. S. Was, P. Ampornrat, G. Gupta, S. Teysseyre, E. A. West, T. R. Allen, K. Sridharan, L. T. Chen, X. Ren, and C. Pister, *J. Nucl. Mater.* **371**, 176 (2007).

126. R. Hui, W. Cook, C. Sun, Y. Xie, P. Yao, J. Miles, R. Olive, J. Li, W. Zheng, and L. Zhang, *Surf. Coat. Technol.* **205**, 3512 (2011).

127. E. Garcia-Verdugo, J. Fraga-Dubreuil, P. A. Hamley, W. B. Thomas, K. Whiston, and M. Polyakoff, *Green Chem.* **7**, 294 (2005).

128. M. Boero, T. Ikeshoji, C. C. Liew, K. Terakura, and M. Parrinello, *J. Am. Chem. Soc.* **126**, 6280 (2004).

129. Y. Ikushima, K. Hatakeda, O. Sato, T. Yokoyama, and M. Arai, *Angew. Chem., Int. Ed.* **40**, 210 (2001).

130. H. Borwick, O. Walter, E. Dinijus, and J. Ribizant, *J. Organomet. Chem.* **570**, 121 (1998).

AUTHOR INDEX

Numbers in parentheses are reference numbers and indicate that the author's work is referred to although his name is not mentioned in the text. Numbers in italics show the page on which the complete references are listed.

Abdulagatov, A., 31(42), *54*
Abdulagatov, A. I., 16(51), *21;* 31(41), *54*
Abdulagatov, A. M., 17(42), *21*
Abdulagatov, I. M., 16(50–51),
    17(43–44), *21;* 26(13), 28(13),
    29(28, 31), 31–32(13), *53;* 31(40–41),
    32(44), *54;* 104(33), 108(56–58),
    109(60–61), 123(115), *145*
Abdurashidova, A., 109(60), *146*
Abe, H., 171(112), 173(112), *181*
Abraham, M. A., 163–164(61), *180*
Abramson, E. H., 26(11), 34(11), *53*
Adschiri, A., 156(27), 157(29), *178*
Adschiri, T., 49(108), *56;* 113(74), 114(78),
    146; 153(11), *178;* 166(74), *180*
Ahonen, A., 164(62), *180*
Aida, T. M., 152(5), *177*
Aidam, T. M., 101(2), *144*
Aizawa, T., 15(27), *20;* 49(108), *56;*
    113(75–77), 114(78), *146;*
    140(181), *150*
Akgün, M., 156(20–21), 157(21),
    161(45), *178;* 161(46), *179*
Akinfiev, N. N., 130(152), *149*
Alben, J., 152(6), 155(6), *177*
Aleksandrov, A. A., 29(29), *53*

Alfantazi, A., 172(119), *182*
Alisultanova, G. S., 16(51), *21;* 31(41), *54*
Allan, N. L., 168(86), *180*
Allen, T. R., 170(125), 173(125), *182*
Allison, L., 170–171(99), *181*
Ally, M. R., 116(88), *147*
Alonso, E., 152(4), 160(4), *177*
Alwani, Z., 107(48–49), *145*
Amend, J. P., 157(28), 168(28), *178;*
    169(94), *181*
Ampornrat, P., 170(125), 173(125), *182*
Anderko, A., 29(26–27), 34(26), *53;*
    117(92–93), 119(92), *147*
Anderson, G. M., 167(81–182), *180*
Anisimov, M. A., 16(33), *20;* 26(116),
    56
Anitescu, G., 140(183), *150;* 162(53),
    179
Anovitz, L. M., 138–139(176), *150*
Antal, Jr., M. J., 155(17), *178*
Antipova, M. L., 72(46), *97*
Arai, K., 15(27), *20;* 49(108), *56;* 72(44),
    97; 101(2), *144;* 113(74, 77–78), *146;*
    131(156), *149;* 140(181), *150;* 152(5),
    177; 153(11), 156(27), *178;* 160(42),
    179; 164(60, 64), *180*

Arai, M., 16(39), *21;* 163(129), *182*
Arai Wei, K., 157(29), *178*
Archer, D. G., 35(53), *54*
Argoud, R., 79(74), *98*
Armellini, F. J., 117(94), 118(94), *147*
Arons, J. de S., 16(35), *20;* 26–27(14), *53;* 83–85(89), *98;* 155(15), *178*
Arslan, T., 156–157(21), *178*
Artemenko, S., 110(65), *146*
Assael, M. J., 40(68), *55*
Asselin, E., 172(119), *182*
Avalos, J. B., 115(82), *147*
Aymonier, C., 101(3), *144;* 151–152(3), *177*
Azadi, P., 15(31), *20*
Azumi, Y., 62(13), *96*
Azzopardi, B., 166(69), *180*

Baes, Jr., C. G. F., 46(93), *55*
Bagchi, B., 109(62), *146*
Balashov, V. N., 123(122), 130(122), *148*
Balbuena, P. B., 128(142), 135(169), *149;* 136–137(172), *150*
Balfour, F. W., 31(38), 33(38), *53*
Bandura, A. V., 46(95), *56*
Baraille, I., 75(55), 77(55), 84(55), *97*
Bass, J. D., 68(30–31), 75(30), 85(30), *96*
Bassez, M.-P., 169(97–98), *181*
Bastea, S., 26(12), 42(12), *53;* 73(47), *97*
Basu, R. S., 16(34, 38), *20;* 39–40(66), 43(85), *55*
Battino, R., 102(8, 10), *144*
Batyrova, R. G., 104(33), *145;* 109(61), *146*
Baur, S., 161(44), *179*
Bazaev, A. R., 17(43), *21;* 29(31), *53;* 108(56–57), 109(60), *146*
Bazaev, E. A., 17(43), *21;* 29(31), *53;* 108(57), *146*
Bazaev, E. E., 109(60), *146*
Bégué, D., 75(55), 77(55), 84(55), *97*
Bekou, E., 40(68), *55*
Bellissent-Funel, M.-C., 8(4), *19,* 42 (75), *55;* 63(16), 65(16, 19–20), *96*

Belyakova, P. E., 30(35), *53*
Bennett, G. E., 49(106), *56;* 111(66), 115(66), *146;* 136–137(172), *150*
Benson, S. W., 129(151), 131(151), *149*
Bergmann, U., 79(74), *98*
Bermejo, M. D., 29(25), 34(25), *53;* 105(38), *145;* 158–160(36), *179*
Bernabei, M., 65(21, 22), 75(21), 89 (21, 22), *96*
Besnard, M., 42(76), *55;* 90(24), *96;* 139(180), *150*
Beta, I. A., 42(75), *55*
Beysens D., 63(16), 65(16), *96*
Bird, R. B., 41(69), *55*
Bischoff, J. L., 117(90–91), *147*
Biswas, R., 109(62), *146*
Blencoe, J. G., 138–139(175–176), *150*
Blood, P., 166(69), *180*
Blumberg, R. L., 60(5), 65(5), 70(5), 88(5), *96*
Bodnar, R. J., 139(177), *150*
Boero, M., 163(128), *182*
Bolander, R. A., 10(10), *19*
Bolanos, G., 161(49), *179*
Bondarenko, G. V., 10(14), *19;* 76(63), 77(64), *98;* 135(168), *149*
Borovlov, A. V., 72(46), *97*
Borwick, H., 163(130), *182*
Botti, A., 63(18), 65(18, 21), 75(21), 89(21), *96;* 106(44), 139(44), *145;* 139(179), *150*
Boughner, E. C., 14(26), *20;* 49(105), *56;* 112(70), 114(70), *146*
Boukis, N., 152(6), 155(6), *177;* 172(115–118), *181*
Bradley, D. J., 11(19), *20*
Brady, P. V., 133(164), *149*
Braunstein, J., 116(88), *147*
Bremholm, M., 15(32), *20*
Brodholt, J. P., 37(59–60), *54;* 69(34), 83 (34), 88(34), *97;* 139(178), *150*
Bröll, D., 163(85), 168(85), *180*
Bröllos, K., 107(50) 108(50), *145*

Brovchenko, I., 69–70(35), 88–89(35), 97; 89(100), 99

Brown, J. M., 26(11), 34(11), 53

Brown, J. S., 14(26), 20; 49(104–105), 56; 112(69–70), 114(70), 128(69), 146

Brown, M. S., 106(45), 140(45), 145

Bruni, F., 63(17–18), 65(17–18, 21), 75(21), 85(17), 89(21), 96; 69(34), 83(34), 88(34), 97; 106(44), 139(44), 145; 139(179), 150

Brunner, E., 107(54), 145; 107(55), 110 (55), 146

Brunner, G., 151(1–2), 158(2), 160(2), 162–163(1–2), 177; 156(23), 178; 158(38), 179

Bryzgalin, O. V., 134(166), 149

Buelow, S., 118(98), 147

Bureau, H., 162(55), 179

Burnham, C. W., 167(81–82), 180

Bury, P., 43(82), 55

Cabani, S., 102(9), 144

Cabet, C., 173(123), 182

Cancell, F., 151–152(3), 177

Canel, M., 15(30), 20

Cannistrato, S., 75(94), 86(94), 99

Cansell, F., 101(3), 144

Canuto, S., 112(67), 146; 115(81), 147

Cao, H., 12(23), 14(23), 20; 46(118), 56

Capman, W. G., 27(18), 53

Carey, D. M., 78(70), 98

Carlon, H. R., 10(17), 20

Casey, C. P., 100(1), 144

Celso, F. Lo, 106(44), 139(44), 145; 139(179), 150

Cezac, P., 153(13), 155(13), 178

Chen, G., 157(30), 178

Chen, L. T., 170(125), 173(125), 182

Chen, X., 12(23), 14(23), 20; 46(118), 56

Cheng, H., 167(79), 180

Cheng, Z.-M., 157(31–132), 178

Chern, S.-M., 25(10), 52

Chialvo, A. A., 133(163), 149

Chlistunoff, J., 131(157), 149

Christoforakos, M., 35(55), 54; 102(21), 144

Chu, T. C., 77(66), 98

Chu, Y. C., 10(15–16), 20; 78(69), 98

Churakov, S. V., 68(32), 96; 71(40–41), 85(40), 97

Clague, D. A., 168(87), 181

Clever, H. L., 102(8), 144

Cocero, M. J., 29(25), 34(25), 53; 105(38), 145; 152(4), 160(4), 177; 158–160(36), 179

Cochran, H. D., 70(37), 97; 104(35), 135(34–35), 145; 133(163), 149

Cohen-Adad, R., 116(86), 147

Concha, M. C., 105(39), 145

Conradi, M. S., 60(3), 75(3), 78(3), 80–81(3), 83(3), 85(3), 95

Contreras, O., 115(82), 147

Cook, W., 173(126), 182

Cooper, J. R., 16(37), 20; 30(33) 32(33), 53; 34(46), 54

Copeland, C. S., 129(151), 131(151), 149

Corradini, M. L., 171(108), 181

Coutinho, K., 112(67), 146; 115(81), 147

Crovetto, R., 102(7), 144

Cui, S. T., 118(96–97), 130(96), 147; 131(160), 135(160), 149

Cummings, P. T., 70(37), 74–75(52), 97; 104(35), 135(34–35), 145; 128(144–146), 133(163), 149

Curtiss, C. F., 41(69), 55

Dahlem, O., 9(46), 21

Dai, Z., 15(29), 20

Danell, A., 102(23), 104(23), 144

Dang, L. X., 72(43), 97

Danten, Y., 42(76), 55; 90(24), 96; 139(180), 150

Darab, J. G., 105(36), 137(36), 145

Davies, G. R., 168(86), 180

de la Ossa, E. J. M., 155(16), 156(16), 178; 161(48), 179

de Loos, Th. W., 107–108(53), 145

DeFries, T., 79(75), 91(75), 98

Dekerckheer, C., 9(46), *21*
Dell'Orco, P., 118(98), *147*
Dell'Orco, P. C., 121(106), *147*
Delmelle, M., 75(94), 86(94), *99*
Dem'yanets, Yu. N., 61(6–8, 10), *96*
Denielou, L., 34(47), *54*
Denyher, J., 166(69), *180*
DePaz, M., 9(45), *21*
Desmarest, Ph., 43(84), *55*
Destrigneville, C. M., 139(178), *150*
DiNero, J. L., 158(39), *179*
Ding, Y., 157(31–32), *178*
Dinius, E., 101(5), *144;* 152(6), 155(6),
  177; 164(60), *180;* 172(115–116),
  181; 163(130), 172(117–118),
  182
DiPippo, M. M., 117(95), *147*
Ditter, W., 76(61), *97*
Dittmann, A., 16(36), *20*
Dixon, C. N., 163–164(61), *180*
Dlott, D. D., 79(73), *98*
Dodd, V. S., 25(9), 33(9), *52*
Donohue, M. D., 27(20–21), *53;* 83(90),
  98; 68(96), 83(96), *99*
Doren, D. J., 130(153), *149*
Dorsey, N. E., 22(4), *52*
Duan, J., 106(43), *145*
Duan, Z., 24(7), 33(7), *52;* 104(32),
  138–139(32), *145*
Dubessy, J., 77(65), *98*
Dudziak, K. H., 39(65), *55*
Dunn, L. A., 123(129), *148*
Dvoryanchikov, V. I., 16(50), *21;* 26(13),
  28(13), 31(39), 31–32(13), *53;*
  123(115), *148*
Dyer, P. J., 74–75(52), *97*
Dyer, R. B., 78(67), *98*

Eaton, H., 118(98), *147*
Eberz, A., 117(89), 123(89), *147*
Eckert, C. A., 15(28), 14(26), *20;*
  49(104–105), *56;* 101(4), *144;*
  112(69–70), 114(70), 128(69), *146*
Economou, I. G., 16(35), *20;* 26(14),
  27(14, 21), *53;* 83(89–90), 84–85(89), *98*

Edwards, H. W., 48(102), *56*
Egorov, B. N., 32(43), *54*
Ehara, K., 156(24–26), *178*
Eisenberg, D., 8–10(7), *19*
Elliot, D. C., 153(7), *177*
Eugster, H. P., 129(150), *149;* 166–168
  (76), *180*

Farnood, R., 15(31), *20*
Fdz-Polanco, F., 152(4), 160(4), *177*
Feng, W., 155(15), *178*
Fernandez, D. P., 11(18), 16(18), *20;*
  35–36(57), 50(57), *54;* 121(109),
  148
Fernandez-Prini, R., 102(7), *144;*
  116(84), *147*
Ferrante, F., 106(44), 139(44), *145;*
  139(179), *150*
Firus, A., 158(38), *179*
Flanagan, L. W., 135(169), *149*
Flowers, G. C., 121(111), 140(111),
  148
Fogo, J. K., 129(151), 131(151), *149*
Fonseca, T. L., 112(67), *146;* 115(81),
  147
Foy, B. R., 78(67), *98*
Fraga-Dubreuil, J., 163(127), *182*
Franck, E. U., 34(51), 34(48), 35(48, 52,
  55), 36(52), 37(52, 63), 44–46(63), *54;*
  39(65), 43(81), 44(88), *55;* 60(4), *95;*
  75–77(53), 85–86(53), *97;* 102
  (11–18, 20–23), 103(11, 16, 24), 104
  (14, 17–20, 22–24), 138(19), *144;*
  104(28–29), 107(28), *145;* 116(87),
  117(87, 89), 123(87, 89), *147;*
  123(131–133, 137), *148;* 107(154),
  131(154, 159, 161), *149*
Frantz, J. D., 77(65), *98;* 123(118),
  124(138), *148*
Fried, L. E., 26(12), 42(12), *53;* 73(47),
  97
Friedman, L., 9(45), *21*
Friend, D. G., 31(40), *54;* 40(68), *55*
Fujisawa, T., 113(72–73), 114(79),
  146

Fujiwara, H., 105(41), *145*
Fulton, J. L., 105(36), 137(36), *145;*
  137(173), *150*

Galkin, A. A., 162–163(58), *179*
Gallagher, J. S., 16(37), *20;* 22(2), 30(2),
  40(2), 43(2), 51(2), *52;* 34(46),
  38–39(64), *54*
Gao, G.-H., 42(77), *55*
Garcia Jarana, M. B., 155(16), 156(16),
  178
Garcia, M. T., 161(48), *179*
Garcia-Verdugo, E., 163(127), *182*
Garrabos, Y., 43(82), *55*
Garrain, P. A., 75(55), 77(55), 84(55), *97*
Gasafi, E., 153(10), *178*
Gasanov, R. K., 108(56–57), *146*
Gebbie, H. A., 10(10), *19*
Geiger, A., 60(5), 65(5), 70(5), 88(5), *96*
George, H. C., 112(67), *146*
Gerber, J., 161(44), *179*
Gianni, P., 102(9), *144*
Giddings, D., 166(69), *180*
Gillespie, S. E., 12(23), 14(23), *20;*
  46(118), *56*
Glaeser, R., 15(28), *20;* 101(4), *144*
Glakounakis, D., 40(68), *55*
Glatzel, P., 79(74), *98*
Gloyna, E., 121(106), *147*
Gloyna, E. F., 131(158), *149*
Goldman, S., 47(98), *56*
Golubev, B. P., 34(50), 13(50), *54*
Goo, G. H., 128–129(149), *149*
Goodwin, A. R. H., 11(18), 16(18), *20;*
  35–36(57), 50(57), *54;* 121(109), *148*
Gorbaty, V., 76(62), *97*
Gorbaty, Yu. E., 10(14), *19;* 57(1), 75(1),
  85(1), *95;* 61(6–8), 62(10–11), 75(11),
  85(11), *96;* 76(63), 77(64), *98;*
  135(168), *149*
Goto, M., 157(34), *178;* 161(43), *179*
Grande, L., 170–171(99), *181*
Gray, C. G., 47(98), *56*
Gray, J. C., 170(100), *181*
Greenwood, H. J., 104(30), 139(30), *145*

Griffith, K., 15(28), *20;* 101(4),
  144
Grigull, U., 22(3), *52*
Gruszkiewicz, M. S., 123(119),
  123–124(125), 132(125), *148*
Guan, Y., 155(18), *178*
Guardia, E., 73(50–51), 74(57),
  83(50–51), 90(57), 91(50–51), *97*
Gubbins, K. E., 27(18), *53*
Guhssani, Y., 63(16), 65(16, 19), *96*
Guillot, B., 63(16), 65(16, 19), *96*
Gülbay, S., 15(30), *20*
Guo, L. J., 155(18), *178*
Guo, S., 157(30), *178*
Gupta, G., 170(125), 173(125), *182*
Gupta, R. B., 26–27(15), 27(17), *53;*
  57(1), 75(1), 85(1), *95;* 83–85(87), *98;*
  86(99), *99;* 135(167), *149*
Guzonas, D., 171(109), *181*

Haar, L., 22(2), 30(2), 40(2), 43(2),
  51(2), *52*
Haglund, P., 162(54), *179*
Hakuta, H., 164–165(65), *180*
Hakuta, Y., 164(64), 165(66), *180*
Hald, P., 15(32), *20*
Halstead, S. J., 47(97), *56*
Hamamoto, T., 114(79), *146*
Haman, S. D., 44(86–87), *55*
Hamley, P. A., 163(127), *182*
Han, M.-H., 42(77), *55*
Hanongbua, S., 71(42), *97*
Harada, M., 14(25), *20;* 81(83), *98*
Harradine, D. M., 78(67), *98*
Harris, J. G., 118(96–97), 130(96), *147;*
  131(160), 135(160), *149*
Hartmann, D., 123(131), *148*
Hartonen, K., 162(51–52), *179;* 164(62),
  180
Harvey, A. H., 16(37), *20;* 17(42), *21;*
  29(28), *53*
Hatakeda, K., 16(39), *21;* 78(68), *98;*
  162–163(59), *179;* 163(129), *182*
Hatano, B., 15(29), *20*
Havey, A. H., 116(84), *147*

Hayashi, H., 164(64), 164–165(65), 165(66), *180*
Hazemann, J.-L., 79(74), *98*
He, F., 156(19), *178*
Heger, K., 34(51), *54*
Hegyi, G. Y., 171(107), *181*
Heikinheimo, L., 171(111), 173(111), *181*
Heilig, M., 102(14), 104(14), *144*
Heinecke, J., 171(106), *181*
Heinzinger, K., 67(28), 70(28), 71(39, 42), 76(39), *96*
Helgeson, H. C., 119(104–105), 121–123(104), 127(104), *147;* 121(108, 110–111), 122(113), 122(112), 123(110, 112, 117), 126 (117), 127(110, 117), 140(111), *148;* 157(28), 168(28), *178;* 167(80), *180;* 168(88–92), 169(93–94), *181*
Hendricks, R. C., 43(85), *55*
Hernandez-Cobos, J., 106(47), *145*
Herzog, H. J., 25(9), 33(9), *52*
Hettiarachchi, S., 171(110), *181*
Hiejima, Y., 37(115), *56;* 82(84–85), 83(84), 91(84–85), *98;* 113(76), *146*
Hill, P. G., 9(8), 14(8), *19*
Hirakoso, H., 157(29), *178*
Hirose, T., 161(43), *179*
Hirschfelder, O., 41(69), *55*
Hnedkovsky, L., 123(122), 130(122), *148*
Ho, P. C., 123(116, 119, 123–124, 127, 130, 134), 132(116, 123, 127, 130, 134), 133(116), *148*
Hodes, M., 160(40, 41), *179*
Hoffman, G. A., 41–42(71), *55*
Hoffmann, M. M., 60(3), 75(3), 78(3), 80–81(3), 83(3), 85(3), *95*
Holmes, H. F., 45(91), *55;* 136–137 (155), 131(155), *149*
Holzapfel, W., 44(88–89), *55*
Honda, T., 38(117), *56*
Hong, G. T., 160(40), *179;* 173(121), *182*
Hordosy, G., 171(107), *181*
Hori, T., 46(96), *56*
Howard, J. B., 158(39), *179*

Hu, H., 157(30), *178*
Hu, Z.-Q., 28(24), *53*
Huang, S. H., 27–28(19), *53;* 84–85(91), *98*
Hui, R., 173(124, 126), *182*
Hyun, J.-K., 128(142–143), *149*

Ibuki, K., 125(140), 127(140–141), 128(140), *149*
Ichikawa, K., 166(74), *180*
Ihm, S.-K., 155(14), *178*
Ikeda, S., 65(25), *96*
Ikeda, Y., 14(25), *20;* 81(83), *98*
Ikeshoji, T., 163(128), *182*
Ikonomou, G. D., 27(20), *53;* 68(96), 83(96), *99*
Ikushima, Y., 16(39), *21;* 37(61), 45(62), 51(62), *54;* 78(68), 83(86), 94(86), *98;* 113(75–76), *146;* 162–163(59), *179;* 163(129), *182*
Inomata, H., 72(44), *97;* 164(60), *180*
Inui, M., 62(13), *96*
Itatani, K., 79(72), *98;* 105(42), *145*
Itoh, K., 65(25), *96*
Itoh, N., 157(35), *179*
Itoh, S., 160(42), *179*
Iversen, B. B., 15(32), *20*
Iversen, S. B., 15(32), *20*
Iwai, T., 171(112), 173(112), *181*
Iwase, H., 65(25), *96*
Izatt, R. M., 12(23), 14(23), *20;* 46(118), *56*
Izrailevskii, L. B., 41(70), *55*

Jackson, G., 27(18), *53*
Jakobs, G. K., 25(8), *52;* 104(31), 120(31), *145*
Jang, J., 173(123), *182*
Japas, M. L., 102(7, 17–18), 104(17–18), *144*
Jay-Gerin, J.-P., 171(109), *181*
Jedlovszky, P., 65–66(23), 69(23), 88–89(23), *96;* 69(34–35), 70(35), 83(34), 88(34–35), 89(100), *97*

Jevremovic, T., 170(103), *181*
Ji, C. M., 155(18), *178*
Jiang, M. H., 165(67), *180*
Jimenez-Espadafor, F., 161(48), *179*
Jockers, R., 107(51–52), *145*
Johnson, D. W., 166(73), *180*
Johnson, J. W., 122(112–113), 123(112), *148*
Johnston, K. P., 26(15), 27(15, 17), *53;* 49(106), *56;* 83–85(87), *98;* 86(99), *99;* 111(66), 112(68), 115(66), *146;* 128(142–143), 131(157–158), 135(167, 169), *149;* 136–137(172), *150*
Jolley, H. R., 123(121, 137), 172–173 (121), *148*
Jonas, J., 41–42(71), *55;* 79(75), 80(76), 91(75), *98*
Jones, E. V., 108(59), 111(59), *146*
Jong, J.-C., 172(120), *182*
Joslin, C. G., 47(98), *56*
Jung, M., 163(85), 168(85), *180*

Kajihara, Y., 37(115), *56;* 62(13), *96;* 82–83(84), 91(84), *98*
Kajimoto, O., 49(107), *56;* 112(71), 113(71), *146*
Kalinichev, A. G., 42(78), *55;* 62(11), 67(28), 68(29–32), 70(28), 75(11, 30), 85(11, 30), *96;* 71(38–41), 76(39), 85(40), *97;* 76(63), *98;* 68(97–98), *99;* 135(168), *149*
Kalpala, J., 162(51), *179*
Kamalov, A. N., 26(13), 28(13), 31–32(13), *53;* 123(115), *148*
Kameda, Y., 65(25), *96*
Kamelov, A. N., 16(50), *21*
Kamgar-Paral, B., 31(38), 33(38), *53;* 23(67), 39(67), 43(85), *55*
Kanakubo, M., 113(75–76), *146*
Kandratsenka, A., 74(58), *97*
Karaborni, S., 70(37), *97;* 104(35), 135 (34–35), *145*
Kato, Y., 38(117), *56*

Kaul, C., 163(85), 168(85), *180*
Kauzmann, W., 8–10(7), *19*
Kawasaki, S.-I., 160(42), *179*
Kazimi, M. S., 171(104–105), *181*
Kell, G. S., 12(47), *21;* 22(2), 30(2), 40 (2), 43(2), 51(2), *52*
Keppler, H., 162(55), *179*
Kerdcharoen, T., 71(42), *97*
Kerrick, D. M., 25(8), *52;* 104(31), 120(31), *145*
Kerschbaum, S., 116(87), 117(87), 123(87), *147*
Khan, M. S., 119(99), *147*
Kharitonov, K. G., 41(70), *55*
Khasanshin, T. S., 29(29), *53*
Kileen, M. A., 40(68), *55*
Kim, H. J., 106(43), *145*
Kim, J., 161(47), *179*
Kim, J.-D., 158(37), *179;* 172(120), *182*
Kim, K., 161(50), *179*
Kim, M.-S., 155(14), *178*
Kim, Y.-C., 161(50), *179*
Kimura, Y., 105(40), *145;* 113(72–73), 114(79), *146*
Kipçak, E., 156(20–21), 157(21), *178*
Kirkham, D. H., 121(111), 140(111), *148*
Kiselev, M., 71(42), *97*
Kiselev, S. B., 16(33), *20;* 17(42), *21;* 29 (28), 31(39), *53;* 31(40), *54;* 26(116), *56*
Kitagawa, K., 153(12), *178*
Kluth, M., 152(6), 155(6), *177*
Knobloch, K., 16(37), *20*
Knox, A. M., 168(91), *181*
Knox, D. E., 101(6), *144*
Koda, S., 79(72), *98;* 105(41–42), *145*
Kodama, A., 161(43), *179*
Kohl, W., 60(4), *95*
Kohno, H., 37(115), *56;* 82–83(84), 91(84), *98*
Konys, J., 173(123), *182*
Korenowski, G. M., 78(70), *98*
Koshizuka, S., 170(102–103), *181*
Kosinski, J. J., 29(27), *53;* 117(93), *147*

Kostyukova, I. G., 16(33), *20;* 26(116), *56*

Krämer, A., 161(44), *179;* 163(85), 168(85), *180*

Krammer, P., 163(85), 168(85), *180*

Kretzschmar, H.-J., 16(36–37), *20;* 34(46), *54*

Krishtal, V, 71(42), *97*

Kritzer, P., 172(113, 115–116), 173(113), 176(113), *181;* 172(117), *182*

Kronholm, J., 162(51–52), *179;* 164(62), *180*

Kruse, A., 101(5), *144;* 152(6), 155(6), *177;* 153(8–9), 154(9), *178;* 164(60), *180*

Kubo, M., 79(71), *98*

Kuge, K., 38(117), *56*

Kumalov, A. N., 31(39), *53*

Kumar, R., 58(2), *95*

Kuosmanen, T., 162(52), *179*

Kurosawa, S., 153(11), *178*

Kusali, J., 156(26), *178*

Kutney, M. C., 25(9), 33(9), *52*

Lamanna, R., 75(94), 86(94), *99*

Lamb, W. J., 41–42(71), *55;* 79(75), 80(76), 91(75), *98*

Landise, R. A., 166(73), *180*

Laria, D., 91(109), *56;* 73(48), 83(48), 91(48), *97*

Larkin, D. K., 29(29), *53*

LaRowe, D. E., 168(92), *181*

Lasdon, L., 131(157), *149*

Laurent, R. L., 168(91), *181*

Lee, I.-G., 155(14), *178*

Lee, S. H., 128(144–149), 129(147, 149), 130(147), *149*

Lee, Y.-W., 161(47), *179;* 166(75), *180*

Lemmon, E. W., 11(18), 16(18), *20;* 35–36(57), 50(57), *54;* 121(109), *148*

Lentz, H., 102(13–15), 104(14), *144*

Lepori, L., 102(9), *144*

Lester, E. D., 166(69), *180*

Letellier, S., 153(13), 155(13), *178*

Leusbrock, I., 119(100–103), 120(102–3), *147*

Levelt Sengers, J. M. H., 11(18), 16(18), *20;* 31(38), 33(38), *53;* 35–36(57), 50(57), 38–39(64), *54;* 121(109), *148*

Leventhal, J. J., 9(45), *21*

Levin, A. Ya., 41(70), *55*

Li, B., 165(66–67), 166(71–72), *180*

Li, J., 173(126), *182*

Li, J.-C., 42(75), *55*

Li, W., 47(98), *56*

Li, Y., 167(79), *180*

Li, Y.-G., 28(24), *53*

Lichtenthaler, R. N., 107–108(53), *145*

Liebscher, A., 167(78), *180*

Liew, C. C., 72(44), *97;* 163(128), *182*

Lilley, M. D., 168(87), *181*

Lindner, H. A., 60(4), *95*

Linton, M., 44(86–87), *55*

Liotta, C. L., 14(26), 15(28), *20;* 49(104–105), *56;* 101(4), *144;* 112(69–70), 114(70), 128(69), *146*

Liu, W., 130(153), *149*

Loewenschuss, A., 136(171), *150*

Loppinet-Serani, A., 101(3), *144;* 151–152(3), *177*

Lu, J., 14(26), *20;* 49(104–105), *56;* 112(69–70), 114(70), 128(69), *146*

Lu, S.-X., 157(31), *178*

Lu, Y. J., 155(18), *178*

Luck, W. A. P., 14(24), *20;* 76(59–61), *97*

Lukashov, Yu. M., 34(50), 13(50), *54*

Lunin, V. V., 162–163(58), *179*

Luo, H., 136(170), *150*

Lvov, S. N., 46(95), *56*

Lyubimov, Yu. A., 11(20–21), 12(20), *20*

Ma, Y., 105(36), 137(36), *145;* 137(173), *150*

Magee, J. W., 17(43, 44), *21;* 29(31), *53;* 31(40), 32(44), *54*

Maltsev, B. K., 31(37), *53*
Mancinelli, R., 106(44), 139(44), *145;* 139(179), *150*
Mangold, K., 131(159), *149*
Mansoori, G. A., 28(22), *53*
Maraczy, C. S., 171(107), *181*
Marcus, Y., 7(3), 8(6), *19;* 11(22), 13–14(22), *20;* 12(49), *21;* 31(36), *53;* 36(58), 40(58), *54;* 22(114), 48(101), 49(114, 119), 52(119), *56;* 84(93), 86–87(93), 88(93, 95), 89(93), *99;* 109 (63), 110(64), *146;* 123–126(114), 130(114), 133(114), 136(114), *148;* 134(165), 136(165), *149;* 136(171), *150*
Mares, R., 16(37), *20*
Marias, F., 153(13), 155(13), *178*
Marrone, P. A., 160(40–41), *179;* 173(121), *182*
Marshall, W. L., 17(41), *21;* 29(30), *53;* 34–35(49), 37(62–63), 38(62), 44–46(63), *54;* 45(91), *55;* 108(59), 111(59), *146;* 123(118, 120–121, 126, 128–129, 135–137), 124(136, 138, 139), 132(120), 172–173(121), *148;* 130(162), 131(155), 136–137(155), *149;* 123(184), *150*
Marti, J., 73(50–51), 83(50–51), 91(50–51), *97*
Martin, A., 29(25), 34(25), *53;* 105(38), *145*
Marulanda, V., 161(49), *179*
Marzari, N., 10(12), *19*
Mason, D. W., 164–165(63), *180*
Masten, D. A., 78(67), *98*
Masters, A. J., 47(97), *56*
Mather, A. E., 102(12), *144*
Matibayasi, N., 41(74), *55;* 81(79–80), 91(79–80), *98*
Matsubayashi, N., 17(40), *21*
Matsuda, K., 62(13), *96*
Matsumura, Y., 155(17), *178*
Matubayashi, N., 41(73), *55;* 65(25), 69(33), *96;* 73(49), 80–81(49), 90–91(49), *97;* 79(71), 80(77–78),

81(81), 83(77), 85(77), *98;* 106(46), *145*
Mausbach, P., 60(5), 65(5), 70(5), 88(5), *96*
Mazur, V., 110(65), *146*
McClain, J. S., 168(87), *181*
McKenzie, W. F., 121(108), *148;* 167(80), *180*
Mesmer, R. E., 45(91), 46(93), *55;* 70(37), *97;* 135(34), *145;* 123(127), 132(127), *148;* 128(144), 133(163), 131(155), 136–137(155), *149*
Metz, S., 119(100–103), 120(102–103), *147*
Meyer, L., 153(10), *178*
Michelberger, Th., 104(28), 107(28), *145*
Mikhael, S., 170–171(99), *181*
Miles, J., 173(126), *182*
Millat, J., 40(68), *55*
Minami, K., 15(27), *20;* 49(108), *56;* 113(74, 77), 114(78), *146;* 140(181), *150;* 166(74), *180*
Misch, B., 158(38), *179*
Miura, S., 42(80), *55;* 72(45), 73(45), 75(45), 85–86(45), 90(45), *97*
Miyafuji, H., 156(22, 25), *178*
Miyagawa, K.16(37), *20*
Miyata, K., 166(74), *180*
Mizan, T. I., 42(79), *55;* 66(27), 71(27), *96*
Mizuta, M., 15(27), *20;* 113(77), *146*
Mizutani, T., 113(74), *146*
Møller, N., 24(7), 33(7), *52;* 104(32), 138–139(32), *145*
Mollica, V., 102(9), *144*
Mori, S., 157(35), *179*
Morimoto, M., 49(103), *56;* 115(83), *147;* 157(33), *178*
Morishita, T., 166(74), *180*
Mountain, R. D., 69–70(36), 75(36), 85(36), *97*
Mulev, Yu. V., 35(54), *54*
Murayama, Y., 38(117), *56*
Murray, J. S., 105(39), *145*
Mursalov, B. A., 31(39), *53;* 123(115), *148*
Mysen, B., 77(65), *98*

Nabokov, O. A., 11(20–21), 12(20), *20*
Nada, T., 161(43), *179*
Nagashima, A., 40(68), *55*
Nakahara, M., 17(40), *21;* 41(73–74),
 55; 65(25), 69(33), *96;* 79(71),
 80(77–78), 81(79–81), 83(77), 85(77),
 91(79–80), *98;* 106(46), *145;* 114(80),
 147; 125(140), 127(140–141),
 128(140), *149*
Nakao, N., 73(49), 80–81(49), 90–91
 (49), *97*
Nakata, T., 156(22, 25), *178*
Nakazomo, Y., 171(112), 173(112),
 181
Nebot Sanz, E., 155(16), 156(16), *178*
Negri, P., 116(86), *147*
Neichel, M., 104(29), *145*
Neindre, B. Le, 34(47), *54;*
 43(82–83), *55*
Neilson, G. W., 63(14–15), 65(15), *96*
Newville, M., 137(173), *150*
Nieto-Draghi, C., 115(82), *147*
Nilsson, A., 79(74), *98*
Nitta, T., 46(96), *56*
Novotny, R., 171(111), 173(111), *181*

O'Shea, S. F., 66(26), 67(26), *96*
Oe, T., 160(42), *179*
Oelkers, E. H., 122(112, 113), 123(112,
 117), 126–127(117), *148*
Ogata, A., 161(43), *179*
Ohashi, K., 49(108), *56;* 114(78), *146*
Ohtaki, H., 62(12), *96*
Ohzono, H., 138(174), *150*
Oka, H., 49(107), *56;* 112(71), 113(71),
 146
Oka, Y., 170(102–103), *181*
Okada, K., 37(61), 45(62), 51(62), *54;*
 37(115), *56;* 82(84), 83(84, 86), 91
 (84), 94(86), *98*
Okada, T., 166(74), *180*
Okazaki, S., 42(80), *55;* 72(45), 73(45),
 75(45), 85–86(45), 90(45), *97*
Okhulkov, A. V., 62(10), *96;* 76(63), *98;*
 135(168), *149*

Okita, N., 16(37), *20*
Oleinikova, A., 69–70(35), 88–89(35),
 97; 89(100), *99*
Olive, R., 173(126), *182*
Olson, E. J., 168(87), *181*
Omori, T., 105(40), *145*
Oparin, R., 139(180), *150*
Orville, P. M., 140(182), *150*
Osada, M., 113(74), *146*
Osawa, K., 113(73), 114(79), *146*
Oscarson, J. L., 12(23), 14(23), *20;*
 46(118), *56*
Oshima, H., 166(74), *180*
Oshima, Y., 166(70), *180*
Otake, K., 114(80), *147*
Otomo, T., 65(25), *96*
Otsu, J., 166(70), *180*
Owens, C. E., 168(91), *181*

Paas, R., 107(51), *145*
Palmer, D. A., 45(91), *55;* 116(84), *147;*
 123(116, 119, 123–124, 127, 130,
 134), 132(116, 123, 127, 130, 134),
 133(116), *148;* 136–137(155), 131
 (155), *149*
Panayiotou, C., 26(16), *53;* 83(88), *98*
Panayiotou, C. G., 26–27(15), 28(23),
 53; 83(87, 92), 84–85(92), 89(92),
 98
Pang, W. S., 165(67), *180*
Pang, Y., 79(73), *98*
Parker, C. M., 168(87), *181*
Parrinello, M., 81(82), *98;* 163(128), *182*
Partay, L. B., 89(100), *99*
Pátay, L., 65–66(23), 69(23), 88–89(23),
 96; 69–70(35), 88–89(35), *97*
Pathania, R., 171(110), *181*
Patrick, H. R., 15(28), *20;* 101(4), *144*
Patterson, J. F., 170(101), *181*
Pei, A. X., 155(18), *178*
Penders, W. G., 107–108(53), *145*
Peng, D. Y., 47(100), *56*
Penttila, S., 171(111), 173(111), *181*
Perrone, R., 116(86), *147*
Peter, K. H., 107(50) 108(50), *145*

Peters, C. J., 16(35), *20;* 26–27(14), *53;* 83–85(89), *98*
Petitet, J. P., 34(47), *54*
Petrenko, V. E., 72(46), *97*
Petrich, G., 152(6), 155(6), *177*
Pettersson, L. G. M., 79(74), *98*
Pfund, D. M., 105(36), 137(36), *145;* 137(173), *150*
Phenix, B. D., 158(39), *179*
Pioro, I., 170–171(99), *181*
Pister, C., 170(125), 173(125), *182*
Pitzer, K. S., 11(19), *20;* 12(48), *21;* 29(26), 34(26), *53;* 35(56), *54;* 46(92), *55;* 117(90, 92), 119(92), *147*
Poirier, J., 173(122), *182*
Pokrovskii, V. A., 122(113), *148*
Poliakoff, M., 166(69), *180*
Politzer, P., 105(39), *145*
Polyakoff, M., 163(127), *182*
Polykhronidi, N. G., 17(44), *21;* 32(44), *54;* 104(33), *145;* 109(61), *146*
Portela, J. R., 155(16), 156(16), *178;* 161(48), *179*
Postorino, P., 63(14–15), 65(15), *96*
Pratt, L. R., 46(94), *56*
Prinos, I., 28(23), *53;* 84–85(92), 89(92), *98*
Pruss, A., 2(1), *19;* 23(6), 31(6), *52*
Ptitsnaya, V. N., 41(70), *55*
Puhovski, Y., 71(42), *97*

Qu, W., 173(124), *182*
Quist, A. S., 34–35(49), *54;* 44(90), 46(90), *55;* 123(120–121, 126, 128, 135–137), 124(136), 132(120), 172–173(121), *148;* 124(139), *149;* 123(184), *150*

Radnai, T., 62(12), *96*
Radosz, M., 27(18–19), 28(19), *53;* 84–85(91), *98*
Ramazanova, A. E., 16(51), 17(43), *21;* 29(31), *53;* 31(41), *54;* 108(56–57), *146*

Räsänen, R., 164(62), *180*
Rasulov, S. M., 108(58), *146*
Reisse, J., 9(46), *21*
Ren, X., 170(125), 173(125), *182*
Rexwinkel, G., 119(100–103), 120(102–103), *147*
Reynolds, T., 118(98), *147*
Ribizant, J., 163(130), *182*
Ricci, M. A., 63(14–15, 17–18), 65(15, 17, 18, 21, 22), 75(21), 85(17), 89(21–22), *96;* 69(34), 83(34), 88(34), *97;* 106(44), 139(44), *145;* 139(179), *150*
Rice, P. A., 162(53), *179*
Rice, S. F., 105(37), *145*
Richard, T., 173(122), *182*
Richter, T., 163(85), 168(85), *180*
Ridard, J., 115(82), *147*
Riekkola, M.-J., 162(51–52), *179;* 164(62), *180*
Rincon, D., 158–160(36), *179*
Ripol-Saragosi, F. B., 34(50), 13(50), *54*
Ritzert, G., 131(161), *149*
Rivkin, S. L., 32(43), *54;* 41(70), *55*
Robinson, D. B. A., 47(100), *56*
Rodriguez-Prado, A., 170–171(99), *181*
Rogak, S., 172(119), *182*
Rogak, S. N., 119(99), *147*
Rollet, A. P., 116(86), *147*
Ropital, F., 172(114), *181*
Rosenzweig, S., 35(55), *54*
Rossky, P. J., 128(142–143), 135(169), *149;* 136–137(172), *150*
Roth, K., 75–77(53), 85–86(53), *97*
Rowlinson, J. S., 10(9), *19*
Rubin, J. B., 48(102), *56*
Ryzhenko, B. N., 134(166), *149*

Sadus, R. J., 102(11–12, 22), 103(11, 24), 104(22, 24), *144*
Saha, P., 171(104–105), *181*
Saito, N., 16(39), *21;* 78(68), *98*
Saito, S., 72(44), *97*
Saitoh, T., 113(75), *146*
Saka, S., 156(22, 24–26), *178*

Sako, K., 117(95), *147*
Sako, T., 114(80), *147*
Samios, J., 74(56), 90(56), *97*
Sanchez, I. C., 26(15–16), 27(15), *53;* 83(87–88), 85(87), *98*
Sanchez-Oneto, J., 155(16), 156(16), *178;* 161(48), *179*
Sandri, E., 152(6), 155(6), *177*
Sanz, M. T., 152(4), 160(4), *177*
Saotome, Y., 157(33), *178*
Sasaki, M., 156(27), 157(34), *178;* 157(35), *179;* 168(84), *180*
Sato, H., 114(79), *146*
Sato, M., 162–163(59), *179*
Sato, O., 163(129), *182*
Sato, S., 49(103), *56;* 115(83), *147;* 157(33), *178*
Sato, T., 153(11), *178;* 157(35), *179;* 164(60), *180*
Satou, W., 46(96), *56*
Saul, A., 23–24(5), *52*
Savage, P. E., 42(79), *55;* 66(27), 71(27), *96;* 162(56–57), 163(57), 164(56), *179*
Sawamoto, S., 157(29), *178*
Scalise, O. H., 103(27), *145*
Schacht, C., 156(23), *178*
Schacht, M., 152(6), 155(6), *177;* 172(118), *182*
Schebek, L., 153(10), *178*
Schiebener, P., 22(3), *52;* 38–39(64), *54*
Schmidt, E., 22(1), *52;* 23(45), 32–33(45), *54*
Schmidt, H., 161(44), *179*
Schmidt, J. R., 58(2), *95*
Schmieder, H., 152(6), 155(6), *177*
Schneider, G. M., 107(48–52), 108(50), *145;* 107(55), 110(55), *146*
Schulenberg, T., 171(106), *181*
Schwarzer, D., 74(58), *97*
Scott, P. L., 103(25), 108(25), 110(25), *144*
Sebastiani, D., 81(82), *98*
Seitz, J. C., 138–139(175–176), *150*
Seminario, J. M., 105(39), *145*

Sengers, J. V., 16(34, 38), *20;* 31(38), 33(38), *53;* 23(67), 39(66–67), 40(66), 43(85), *55*
Serin, J. P., 153(13), 155(13), *178*
Seward, T. M., 102(16), 103(16), *144*
Shim, Y., 106(43), *145*
Shin, N. C., 161(47), *179*
Shin, Y. H., 161(47), *179*
Shiraishi, T., 157(34), *178*
Shnonov, V. M., 102(22), 104(22), *144*
Shock, E. L., 119(104–105), 121–123 (104), 127(104), *147;* 122(112–113), 123(112), *148;* 168(89–90), *181*
Simoneit, B. R. T., 169(95–96), *181*
Simonson, J. M., 17(41), *21;* 45(91), *55;* 70(37), *97;* 135(34), *145;* 128(144), 131(155), 133(163), 136–137(155), *149*
Simonson, J. W., 29(30), *53*
Sinag, A., 15(30), *20*
Sirota, A. M., 30(35), 31(37), *53;* 31 (113), *56*
Sit, P. L.-L., 10(12), *19*
Skaf, M. S., 91(109), *56;* 73(48), 83(48), 91(48), *97*
Skarmoutsos, I., 74(56–57), 90(56–57), *97*
Skinner, J. L., 58(2), *95*
Slanina, Z., 10(11, 13), *19*
Smirnov, S. N., 35(54), *54*
Smith, Jr., R. L., 113(75), *146;* 152(5), *177;* 153(11), *178;* 164(60), *180*
Smith, K. A., 25(9), 33(9), *52;* 158(39), 160(40–41), *179*
Smith, R. D., 164–165(63), *180*
Smith, Jr., R. L., 101(2), *144;* 113(76), *146;* 140(181), *150*
Smits, P. J., 16(35), *20;* 26–27(14), *53;* 83–85(89), *98*
Soave, G., 47(99), *56*
Sögüt, O. Ö., 156(20–21), 157(21), *178;* 161(45–46), *179*
Son, S. H., 161(50), *179*

Soper, A. K., 63(14–15, 17–18), 65(15, 17–18, 21), 75(21), 85(17), 89(21), *96;* 69(34), 83(34), 88(34), *97;* 106(44), 139(44), *145;* 139(179), *150*

Sorai, M., 168(84), *180*

Span, R., 16(37), *20*

Sretenskaja, N. G., 102(11), 103(11), *144*

Sridharan, K., 170(125), 173(125), *182*

Stanley, H. E., 60(5), 65(5), 70(5), 88(5), *96*

Starflinger, J., 171(106), *181*

Steeper, R. R., 106(45), 140(45), *145*

Steinberg, J., 155(17), *178*

Stepanov, G. V., 16(51), 17(44), *21;* 31(41–42), 32(44), *54;* 104(33), *145;* 109(61), *146*

Stern, E. Q. A., 137(173), *150*

Sterner, S. M., 139(177), *150*

Stocker, I., 16(36–37), *20*

Straub, J., 22(3), *52;* 38–39(64), *54*

Sue, K., 131(156), *149;* 140(181), *150;* 160(42), *179*

Sugeta, T., 114(80), *147*

Sugimoto, K., 105(41), *145*

Sugimoto, N., 113(75), *146*

Sugimoto, Y., 157(33), *178*

Sun, C., 173(124, 126), *182*

Sungt, G., 128–129(149), *149*

Suzuki, A., 160(42), *179*

Suzuki, M., 15(27), *20;* 49(108), *56;* 113(77), 114(78), *146*

Suzuki, T., 140(181), *150*

Sverjensky, D. A., 122(112–113), 123(112), *148;* 168(90), *181*

Svishchev, I. M., 166(68), *180*

Sweeton, F. H., 46(93), *55*

Swiata-Wojcik, D., 74–75(54), 84(54), *97*

Syed, K. M., 15(31), *20*

Szala-Bilnik, J., 74–75(54), 84(54), *97*

Tagaya, H., 15(29), *20*

Tagirov, B. R., 130(152), *149*

Takahashi, H., 46(96), *56*

Takami, S., 166(74), *180*

Takanohashi, T., 49(103), *56;* 115(83), *147;* 157(33), *178*

Takebayashi, Y., 114(80), *147*

Tamura, K., 62(13), *96*

Tang, H., 153(12), *178*

Tanger, IV, J. C., 46(92), *55*

Tanger, J. C., 121(110) 123(110), 127(110), *148*

Tanger, Jr., J. C., 12(48), *21*

Tao, X. T., 165(67), 166(71–72), *180*

Tassaing, T., 63(16), 65(16, 19), 90(24), *96;* 75(55), 77(55), 84(55), *97;* 42(76), *55;* 139(180), *150*

Tavlarides, L. L., 140(183), *150;* 162(53), *179*

Tawa, G. J., 46(94), *56*

Temesvary, E., 171(107), *181*

Temkin, M., 103(26), *144*

Terakura, K., 163(128), *182*

Terazima, M., 113(72–73, 79), *146*

Testemale, T., 79(74), *98*

Tester, J. W., 25(9), 33(9), *52;* 117(94–95), 118(94), *147;* 158(39), 160(40–41), *179*

Teysseyre, S., 170(125), 173(125), *182*

Thies, M. C., 107(55), 110(55), *146*

Thomas, W. B., 163(127), *182*

Tödheide, K., 102(20, 23), 104(19–20, 23), 138(19), *144*

Toivonen, A., 171(111), 173(111), *181*

Tomberli, B., 47(98), *56*

Tomiyasu, H., 14(25), *20;* 81(83), *98*

Tong, Q.-Y., 42(77), *55*

Tortorelli, P. F., 173(123), *182*

Touba, H., 28(22), *53*

Toyoshima, K., 113(74), *146*

Tranquard, A., 116(86), *147*

Tremaine, P., 171(109), *181*

Tremaine, P. R., 66(26), 67(26), *96*

Triolo, R., 106(44), 139(44), *145;* 139(179), *150*

Tromp, R. H., 63(14–15), 65(15), *96*

Trubenach, J., 16(36), *20*

Tsederberg, N. V., 29(29), *53*

Tsukahara, T., 14(25), *20;* 81(83), *98*
Tucker, S. C., 136(170), *150*
Tufeu, R., 34(47), *54;* 43(82–84), *55*

Uchida, H., 79(72), *98;* 105(42), *145*
Uematsu, M., 34(51), 35–37(52), *54*
Ueno, M., 125(140), 127(140–141), 128(140), *149;* 166(74), *180*
Ungerer, P., 115(82), *147*
Uosaki, Y., 41(74), *55;* 81(80), 91(80), *98*
Urisova, M., 116(85), *147*
Uskan, B., 15(30), *20*

Vadillo, V., 161(48), *179*
Vallauri, R., 69(34), 83(34), 88(34), *97*
Valyashko, V., 116(85), *147*
Valyashko, V. M., 116(84), *147*
van der Kooi, H. J., 155(15), *178*
van Konynenburg, P. H., 103(25), 108(25), 110(25), *144*
van Sijl, J., 168(86), *180*
van Westrenen, W., 168(86), *180*
Vargaftig, N. B., 30(34), *53*
Vastamäki, P., 164(62), *180*
Vazquez, V., 158–160(36), *179*
Ved, O. V., 72(46), *97*
Vega, L. F., 106(47), *145*
Vera, J. H., 28(23), *53;* 84–85(92), 89(92), *98*
Veriansyah, B., 158(37), 161(47), *179;* 172(120), *182*
Versteeg, G. F., 119(100–103), 120(102–103), *147*
Villamere, B., 170–171(99), *181*
Villanueva, J. A. B., 161(48), *179*
Vlachou, T., 28(23), *53;* 84–85(92), 89(92), *98*
Vogel, H., 163(85), 168(85), *180*
Vöhringer, P., 74(58), *97*
Von Damm, K. L., 168(87), *181*

Wagner, W., 2(1), 11(1), *19;* 16(37), *20;* 23(5–6), 24(5), 31(6), *52*
Wahyudiono, 157(34), *178*

Wakai, C., 17(40), *21;* 69(33), *96*
Wakai, Ch, 41(73), *55;* 80(77–78), 81(81), 83(77), 85(77), *98*
Wakita, H., 61(9), *96;* 138(174), *150*
Wallen, S. L., 137(173), *150*
Walrafen, G. E., 10(15–16), *20;* 77(66), 78(69), *98*
Walter, O., 163(130), *182*
Walther, J. V., 121(107), 140(107), *147;* 133(164), *149;* 140(182), *150;* 166–167(77), 168(83), *180*
Wang, P., 35(53), *54*
Wang, Z., 79(73), *98*
Was, G. S., 170(125), 173(125), *182*
Wasserman, E., 37(59–60), *54*
Watanabe, M., 113(74), *146;* 157(29), *178;* 157(35), *179;* 164(60), *180*
Watson, J. T. R., 16(34), *20;* 39–40(66), 43(85), *55*
Weare, J. H., 24(7), 33(7), *52;* 104(32), 138–139(32), *145*
Weber, I., 16(37), *20*
Wei, C., 156(19), *178*
Wei, Y. S., 103(24), 104(24), *144*
Weingärtner, H., 8(5), *19*
Welsch, H., 102(15), *144*
Wernet, Ph., 79(74), *98*
West, E. A., 170(125), 173(125), *182*
Whalley, E., 12(47), *21*
Whalley, R. D. C., 9(8), 14(8), *19*
Whiston, K., 163(127), *182*
Wickham, J. J., 105(37), *145*
Wilcock, R. J., 102(10), *144*
Wilhelm, E., 41(72), *55;* 102(10), *144*
Williams, L. L., 48(102), *56*
Williams, R. C., 11(18), 16(18), *20;* 35–36(57), 50(57), *54;* 121(109), *148*
Willkommen, T., 16(36), *20*
Wilson, J. A., 171(110), *181*
Wofford, W. T., 121(106), *147;* 131(158), *149*
Wood, B., 37(59–60), *54*
Wood, R. H., 123(122), 130(122), *148;* 130(153), *149*
Wood, R. J., 139(178), *150*

Wood, T. H., 123–124(125), 132(125), *148*
Woodland, A. B., 168(83), *180*
Woolley, H. W., 30(32), 32(32), *53*
Wright, J. H., 170(100–101), *181*
Wu, G., 102(14), 104(14), *144*
Wu, J., 156(19), *178*

Xiang, T., 112(68), *146;* 131(158), *149*
Xie, C., 156(19), *178*
Xie, Y., 173(126), *182*
Xiong, P., 167(79), *180*
Xu, T., 167(79), *180*
Xu, X., 155(17), *178*
Xue, M., 165(67), *180*

Yalkowsky, S., 7 (2), *19*
Yamagami, M., 138(174), *150*
Yamagata, T., 166(74), *180*
Yamaguchi, T., 61(9), 62(12), *96;* 138(174), *150*
Yamanaka, K., 61(9), *96*
Yamasaki, T., 170(102), *181*
Yan, B., 156(19), *178*
Yang, J.-C., 28(24), *53*
Yang, W.-H., 10(15–16), *20;* 78(69), *98*
Yao, M., 37(61), 45(62), 51(62), *54;* 37(115), *56;* 82(84–85), 83(84–86), 91(84–85), 94(86), *98*
Yao, P., 173(126), *182*
Yasaka, Y., 79(71), *98*
Yick, S., 173(124), *182*
Yling, T., 104(28), 107(28), *145*
Yoda, S., 114(80), *147*
Yokoyama, T., 163(129), *182*

Yoshida, K., 41(73–74), *55;* 81(79–80), 91(79–80), *98;* 138(174), *150;* 156(26), *178*
Yoshie, H., 42(80), *55;* 72(45), 73(45), 75(45), 85–86(45), 90(45), *97*
Yoshii, N., 42(80), *55;* 72(45), 73(45), 75(45), 85–86(45), 90(45), *97*
Yoshizawa, Y., 38(117), *56*
Yu, Y.-X., 42(77), *55*
Yuan, P.-Q., 157(31–32), *178*
Yuan, W.-K., 157(31–32), *178*
Yui, K., 79(72), *98;* 105(42), *145*

Zehner, P., 163(85), 168(85), *180*
Zetzl, C., 156(23), *178*
Zhan, Y. J., 165(67), *180*
Zhang, H. D., 165(67), *180*
Zhang, J., 157(30), *178*
Zhang, L., 173(126), *182*
Zhang, R., 162–163(59), *179*
Zhang, W., 167(79), *180*
Zhang, X. M., 155(18), *178*
Zhao, H., 63(16), 65(16), *96*
Zhao, J., 171(104–105), *181*
Zhao, L.-Q., 157(31–32), *178*
Zheng, Q. M., 166(71), *180*
Zheng, Q. X., 165(67), 166(72), *180*
Zheng, W., 173(126), *182*
Zheng, Y., 167(79), *180*
Zhou, W., 162(53), *179*
Ziegler, K. J., 131(157), *149*
Zierenberg, A. A., 168(87), *181*
Ziff, R. M., 42(79), *55;* 66(27), 71(27), *96*
Zotov, A. V., 130(152), *149*

SUBJECT INDEX

Acetone   109, 111, 114, 115
Air   102, 104, 105, 142, 158, 160, 162, 175
Alkali metal nitrate   118, 119, 120, 123
Alumina   165–167, 172, 173, 177
Antisolvent   6, 151
Argon   102–105, 141

Benzene   107, 108, 110, 115
Benzophenone   111, 112, 142
BET   116, 166
Biofuel   152, 174
Biomass   152, 153, 156, 174
Biomolecules   169, 176
Bitumen   157, 174

Calcite   140
Calcium chloride   120, 122, 124, 129
Carbon dioxide, as solute   102, 104, 106, 138–142, 152–156, 166–169, 174
Carbon dioxide, supercritical (SCD)   6, 8, 100, 138, 143, 162, 164
Carbon monoxide   152–155, 160, 174
Catalysts   153, 156, 163, 164, 169, 174
Cavity formation   115
Ceramics   172, 177
Chemical oxygen demand (COD)   155, 156, 158, 161, 175
Clusters   9, 58, 71, 79, 83, 90, 94, 111, 113, 139, 140, 142

Cohesive energy density   11
Collective motions   77, 78
Compressibility   4, 23, 32, 34, 114
Compressibility factor   8, 18, 23, 24, 50, 104
Computer simulation, molecular dynamics   37, 42, 47, 57, 66, 70–74, 83–86, 90–93, 95, 104–106, 115, 118, 128, 129, 133, 135, 139, 143, 168
Computer simulation, Monte Carlo   57, 66–70, 85, 88, 92, 112, 114, 115, 139
Conductivity, electrical   37, 38, 51
Conductivity, thermal   16, 42, 43, 51
Conformal fluids   7, 23
Coordination number   11, 13, 58, 62
Corrosion problems   101, 141, 152, 158, 171–174, 177
Critical density   2, 23
Critical enhancement   16, 39, 43
Critical indices   4, 5, 15–17
Critical pressure   2–4, 17, 23
Critical temperature   2, 4, 23, 67, 71, 72, 90, 93

Degrees of freedom   1, 2, 49
Dielectric relaxation spectroscopy   37, 73, 82, 83, 91, 95
Dipole orientation parameter   11, 13, 35, 36, 40, 50, 88

Supercritical Water A Green Solvent: Properties and Uses, First Edition. Yizhak Marcus.
© 2012 John Wiley & Sons, Inc. Published 2012 by John Wiley & Sons, Inc.

Electrostriction 136, 137
Energy balance 152, 158–161, 174, 175
Enthalpy, molar 32, 33, 50
Entropy, molar 32, 33, 50
EoS model SAFT 27, 28, 42, 84, 85
EoS model, APACT 16, 27, 83, 85
EoS model, LHFB 26–28, 83, 85
EoS, cubic 25, 47, 48, 110, 142, 155
EoS, scaling equations 26–29, 32, 105, 119, 121, 142
Equations of State (EoS) 22–29, 49, 50, 155
Ethanol 109, 156, 174

Gases, solubility of 101–104, 141
Gas-gas immiscibility 102, 108, 141
Geothermal aspects 166–169, 176

H-bond lifetime 57, 74, 90, 91, 95, 115
H-bond signature peak 63, 65, 68, 72, 92, 93
H-bonding criteria 58, 68–74, 89–92, 94
H-bonds per water molecule 59, 60, 68, 70–75, 77, 81, 85–88, 92–94, 114, 135, 139
Heat capacity, isobaric 30–32, 50
Heat capacity, isochoric 16, 31, 32, 50, 87
Heavy water (D$_2$O) 9, 14, 17, 29, 32, 41, 107, 138
Helmholz energy, molar 34, 117
Hexane 107, 108, 142, 160
Hydrocarbons 107, 108, 142, 176
Hydrochloric acid 123, 129–131, 134, 136, 137, 143, 158, 167, 172
Hydrogen 102, 103, 105, 141, 151–156, 162, 169, 173, 174, 177
Hydrogen peroxide 158, 161, 162, 175
Hydrogen sulfide 155, 168, 169, 174
Hydrogenation 163, 175
Hydrolysis 44, 163–165, 168, 175, 176
Hydrothermal vents 168, 169

IAPWS 2, 16, 23, 39
Infrared spectroscopy 10, 74–77, 84, 85, 93, 139
Internal pressure 11, 18
Ion (salt) conductivity 44, 123–129, 143
Ion association (ion pairing) 116, 118, 127, 129–134, 142, 143
Ion diffusion 127, 128
Ion hydration 116, 118, 121, 128, 134–138, 1042, 143
Ionic radius 122, 128, 134
Ions, hydration numbers of 118, 135, 143
Ions, thermodynamic functions of 122, 123, 136, 137, 143, 176

Law of rectilinear diameters 5, 15, 17
Linear solvation energy relationship (LSER) 7, 17
Lithium chloride 123, 125, 128, 131, 134, 138
Local density 74, 93, 111–114, 135, 136, 142

Magnesium chloride 120, 124, 129
Metal oxides 164, 165, 175
Methane 102, 104, 106, 152–156, 164, 168, 169, 174
Methanol 109, 140, 143, 162, 164
Mineral acids, solubility of 116, 142
Mixtures of SCD with SCW 138–140, 143, 167, 168, 176
Monomeric water 26, 27, 50, 60, 68–71, 74–77, 79, 80, 83, 84, 93–95

Near critical water 15–17, 19, 26, 101, 112, 177
Neutron diffraction 42, 57, 61, 62, 83, 85, 90, 92, 138, 139
Nickel-base alloys 172, 173, 177
Nitric acid 131, 134

Nitrogen   102, 104, 105, 142, 155,
 161, 169
NMR chemical shift   14, 17, 19, 69, 80,
 81, 85, 87, 94, 114, 115
Nuclear magnetic resonance
 (NMR)   78–81, 85, 91
Nuclear power reactors   169–171,
 173, 176

O–H bond stretching   10, 14, 74, 76–79,
 93, 105, 106
Oswald coefficient   101, 141
Oxygen   102, 104, 105, 142, 155–158,
 161, 163, 173, 175

Pair correlation function   58–61,
 67–72, 92, 104
PCHs, PAHs, dioxins, etc. 110, 140, 158,
 161, 162
Percolation   60, 68–72, 83, 89–93, 106,
 139, 142
Permittivity, relative   12, 13, 18, 34–37,
 50, 115, 121, 122, 134, 166
Petroleum   169, 176
Phase diagram   1–3, 17, 102–104,
 106–111, 116
Polar organic solutes   108, 109
Potassium hydroxide   120, 121, 122,
 123, 131, 155, 156, 160, 167, 175
Potential model (BJH, SPC/E,
 TIP4P etc)   67–71, 73, 83, 93,
 106
Powder technology   164–166, 175
PVT data   22, 23, 49, 86, 87
Pyrolysis   157

Quasilattice quasichemical model   109

Radial distribution function   57, 61,
 67, 74
Radiolysis   171, 173
Raman spectroscopy   10, 16, 77–79,
 105, 106, 113, 114, 139, 140
Reduced variables   4, 17, 24
Refractive index   38, 39

Relaxation, re-orientation time   11, 73,
 744, 79–83, 90, 91, 95, 106
RESS technique   6, 151, 152, 163–164,
 169, 174

Salts, solubility of   115–121, 142
Saturation line   2–4, 11, 14, 17
Scaled particle theory   106
SCWG process   152–157, 172–174,
 177
SCWO process   116, 117, 157–162,
 172–175, 177
Sea water   168
Self-diffusion coefficient   41, 42, 51,
 66, 71, 72, 81, 87, 93
Silica   121, 143, 165–168, 176
Sodium carbonate   119, 120, 158
Sodium chloride   117, 118, 120, 122,
 123, 125, 127–131, 133–135, 142,
 160, 161, 166, 167
Sodium hydroxide   116, 117, 123–126,
 130, 131, 133, 136, 140, 142, 143, 158,
 164, 167
Sodium sulfate   117, 119, 120,
 123, 130
Solubility parameter (Hildebrand)   7,
 47, 52, 110, 142
Solubility parameter, Hansen   48, 49,
 52, 115
Solvatochromic probes, parameters   7,
 14, 15, 19, 49, 111–114, 140, 142
Sound velocity   32, 34
Stainless steel   172, 173, 177
Steam tables   11, 22, 39, 43
Structure, structuredness   11, 13, 14,
 17, 18, 31
Sulfuric acid   123, 130, 134, 158, 172
Supercritical fluid (SCF)   3–7, 17, 100,
 101
Syn-gas   152, 155, 174
Synthetic reactions   162–164, 175

Tetrahedral water   58, 62, 68
Total organic carbon (TOC)   156, 158,
 160–162, 175

Vapor-liquid equilibria (VLE)   2, 17, 22
Virial coefficients   9, 101, 108
Viscosity   16, 18, 39–41, 51, 128
Volume, molar   4, 23, 86, 102, 121, 143, 168

Waste water, sewage   152–156, 160, 161, 174
Water dimers   10, 18, 139, 169
Water gas shift reaction   153, 174

Water ion product   14, 37, 38, 44–47, 51
Water, gaseous   8–11, 30, 32, 50
Water, liquid   11
Water, molecular properties   9
Water, polarizability   13, 35, 3.6, 39

X-ray absorption fine structure (XAFS)   105, 137, 142
X-ray diffraction   57, 60–62, 92
X-ray Raman scattering   79, 94

Printed and bound by CPI Group (UK) Ltd, Croydon, CR0 4YY

16/04/2025

14658343-0004